Markov Model - Theory and Applications

Edited by Hammad Khalil and Abuzar Ghaffari

Published in London, United Kingdom

Markov Model - Theory and Applications
http://dx.doi.org/10.5772/intechopen.1001541
Edited by Hammad Khalil and Abuzar Ghaffari

Contributors
Takashi Honda, Yukiko Iwata, Lyes Ikhlef, Sivasamy Ramasamy, Wilford B. Molefe, Omar Thomas, John Sobanjo, Doncho S. Donchev, Patricia Martinez Vara, Juan Carlos Atenco Cuautle, Elizabeth Saldivia Gomez, Gabriel Martinez Niconoff, Majed M. Alwateer, Mahmoud Elmezain, Mohammed Farsi, Elsayed Atlam

First published in London, United Kingdom, 2023 by IntechOpen
IntechOpen is the global imprint of INTECHOPEN LIMITED, registered in England and Wales, registration number: 11086078, 5 Princes Gate Court, London, SW7 2QJ, United Kingdom

British Library Cataloguing-in-Publication Data
A catalogue record for this book is available from the British Library

Additional hard and PDF copies can be obtained from orders@intechopen.com

Markov Model - Theory and Applications
Edited by Hammad Khalil and Abuzar Ghaffari
p. cm.
Print ISBN 978-1-83769-721-2
Online ISBN 978-1-83769-720-5
eBook (PDF) ISBN 978-1-83769-722-9

Meet the editors

Dr. Hammad Khalil is an accomplished mathematician with a strong passion for knowledge. He earned his BS degree in 2010 and later completed a Ph.D. in Applied Mathematics in 2016. Dr. Khalil's academic journey has made a significant impact on the fields of mathematics and philosophy. He progressed from an assistant professor to an associate professor, leveraging his expertise in scientific computations and programming languages to explore scientific and cosmic mysteries. Dr. Khalil is recognized for his pioneering research and has published multiple papers in ISI-listed journals. His contributions go beyond traditional boundaries, shedding light on the beauty of mathematics and its profound philosophical implications. Additionally, he serves as an editorial board member for respected journals, playing a role in shaping scientific discourse and inspiring others to appreciate the wonders of learning and knowledge.

Dr. Abuzar Ghaffari is a distinguished figure in the realm of fluid dynamics. He holds a BSc and Ph.D. in applied mathematics and has dedicated his academic journey to unraveling the complexities of fluid behavior. Dr. Ghaffari's prolific contributions extend beyond research, encompassing a wealth of mathematics papers that illuminate various facets of the field. Alongside his scholarly pursuits, he has proven to be an exceptional educator in applied mathematics, nurturing a deep understanding of the subject in his students. Dr. Ghaffari's unwavering passion for both exploration and instruction enriches the world of fluid dynamics and applied mathematics.

Contents

Preface

Within the captivating pages of this book, we embark on an illuminating journey into the fascinating world of mathematics and its diverse applications. Here, the collective expertise of renowned scholars and researchers converge to push the boundaries of mathematical theory and practical implementation.

The voyage commences with Chapter 1 and a deep dive into "Markov Operators on a Banach Lattice and Their Applications". Unveiling the elegance of Markov operators, this chapter reveals their significance in the realm of Banach lattices. As the foundation of our exploration, Markov's theory sets the stage for the exciting revelations that follow.

Next, we navigate the intricacies of queuing theory and its practical applications in Chapter 2, "Performance Modeling of Finite Sources Retrial Queue using Markov Regenerative Approach". Delving into performance modeling with a regenerative approach, we uncover the art of predicting and optimizing queuing systems.

The journey continues with Chapter 3, "Markov Models Related to Inventory Systems, Control Charts, and Forecasting Methods". Here, the power of Markov models becomes apparent as we explore their applications in inventory management and control chart analysis. The dynamic interplay of mathematics and real-world challenges takes center stage.

In Chapter 4, "The Application of Markov and Semi-Markov Models in Transportation Infrastructure Management", we venture into the world of transportation systems. Decision-making processes are revolutionized through the insightful utilization of Markov and Semi-Markov models, reshaping the future of infrastructure management.

Navigating decision-making under uncertainty, Chapter 5, "Some Applications of Partially Observable Markov Decision Processes", challenges us with complex scenarios. This chapter unravels the intriguing world of Partially Observable Markov Decision Processes, offering invaluable insights for tackling the unknown.

The wonders of optics and stochasticity come to life in Chapter 6, "Engineering Stochastic Optical Modes". Discovering the ingenuity of engineering optical modes, we witness the brilliance of mathematical application in the domain of optics.

Lastly, our journey concludes with Chapter 7, "Hidden Markov Models for Pattern Recognition". Here, the power of Hidden Markov Models is harnessed to unravel complex data patterns. Through pattern recognition, we gain new perspectives and solutions to intricate challenges.

Collectively, these chapters epitomize the brilliance and versatility of mathematics, showcasing its profound impact on a multitude of disciplines. The authors' dedication and groundbreaking research exemplify the unwavering spirit of exploration that drives progress in the scientific community.

As readers venture through the depths of each chapter, they will gain a deeper appreciation for the ingenuity and relevance of mathematical theory. Whether mathematicians, scientists, or curious minds, this book serves as an invaluable resource, inspiring all to embrace the beauty of knowledge and its boundless applications.

Let the journey begin, and may the revelations within these pages ignite the spirit of discovery, innovation, and collaboration among those who seek to unravel the mysteries of the mathematical world.

Hammad Khalil
Associate Professor,
Department of Mathematics,
University of Education,
Lahore, Pakistan

Dr. Abuzar Ghaffari
University of Education,
Attock Campus,
Attock, Pakistan

Chapter 1

Markov Operators on a Banach Lattice and Their Applications

Takashi Honda and Yukiko Iwata

Abstract

Any Markov process is defined by a stochastic kernel. By using a stochastic kernel, we can define a Markov operator on a Banach lattice. Conversely, by using a Markov operator, we can define a Markov process. There is a close relationship between a Markov process and a Markov operator. In this chapter, we show the relation between a Markov process and a Markov operator. We show some applications of Markov operators concerned with dynamical systems and we mention the Jacobs-de Leeuw-Glicksberg decomposition which is induced by a Markov operator.

Keywords: Markov operators, ergodic theorems, spectral theory, Banach lattice, Markov process

1. Introduction

We shall consider a random dynamical system. Let $S : \mathbb{R} \to \mathbb{R}$ be a dynamical system and define a stochastic process, $\{X_n\}_{n \geq 0}$ by $X_{n+1} = S(X_n) + Y_n$, where $Y_0, Y_1, \cdots$ are independent random variables with values in $\mathbb{R}$, each having the same density g, and X_0 and $\{Y_n\}_{n \geq 0}$ are independent. Then the stochastic process $\{X_n\}$ is a Markov process. Let f_n be the density function of X_n for each $n \geq 0$, and hence, we have $f_{n+1}(x) = \int_{\mathbb{R}} f_n(y)g(x - S(y))\mu(dy)$, where μ is the Lebesgue measure on $\mathbb{R}$. This equation means that every density function f_n is represented by a Markov operator, $T : L^1(\mathbb{R}) \to L^1(\mathbb{R})$ defined by: $Tf(x) = \int_{\mathbb{R}} f(y)g(x - S(y))\mu(dy)$ as $f_n = T^n f_0$. In this way, there are many results about the asymptotic behavior of their distributions represented by positive linear contractions, often called Markov operators. A positive linear contraction on L^1 space is associated with a transition probability for a Markov process. The iteration of a positive linear contraction expresses the asymptotic behavior of a Markov process. Especially, probability density functions with distributions for some Markov processes induced by dynamical systems and noises are sometimes represented by a Markov operator. In this work, we will explain our previous work such as some relations between the Jacobs-de Leeuw-Glicksberg decomposition of semigroups and the existence of a constrictor, which attracts densities of the Markov operators on a Banach space, which are induced by stochastic difference equations. Jacobs first obtained this splitting theorem under the reflexivity assumption. De Leeuw and Glicksberg showed the splitting theorem for an abelian weakly almost periodic semigroup of bounded operators on a complex Banach space. They also showed a similar splitting theorem for a non-abelian semigroup of linear contractions in a strictly convex Banach space with a strictly convex

IntechOpen

dual space. The Jacobs-de Leeuw-Glicksberg decomposition holds for a complex Banach space. We introduce our recent results about some relations of the existence of a constrictor of a linear contractive operator between (continuous or discrete) semigroups of linear contractive operators in a complex Banach space by using our results. Furthermore, we will consider a Markov operator T on a real $L^1(\Omega, \Sigma, \mu)$ space, where (Ω, Σ, μ) is a probability measure space. The iteration of a positive linear contraction expresses the asymptotic behavior of a Markov process. Especially, probability density functions with distributions for some Markov processes induced by dynamical systems and noises are sometimes represented by a Markov operator. We study their asymptotic phenomena by using mean ergodic theorems and pointwise ergodic theorems. In this work, we study our recent operator theoretic approaches for Markov processes and their applications.

2. An operator theoretic approach to a Markov process

A stochastic kernel is a map that in the general theory of Markov processes plays the role that the transition matrix does in the theory of Markov processes with a finite state space. We shall give the basic definition concerning a stochastic kernel (see [1–6]).

Definition 1. A *stochastic kernel* is any map $K : E \times \mathscr{E} \to [0, 1]$ satisfying the following three conditions:

1. for any fixed set $A \in \mathscr{E}$, the function $K(\cdot, A)$ is measurable;

2. for any fixed state $x \in E$, the set function $K(x, \cdot)$ is a measure on $(E, \mathscr{E})$;

3. for all $x \in E, K(x, E) = 1$.

Firstly, we show that any positive contraction on L^1-space induces the stochastic kernel. Let E be a nonempty set, $\mathscr{E}$ a σ-algebra of subsets of E, and $(E, \mathscr{E}, m)$6a σ-finite measure space. Let $L^1(E, \mathscr{E}, m) = L^1$ be the space of real valued integrable functions with respect to $(E, \mathscr{E}, m)$ and $\langle x^*, x \rangle$ be a dual pair of $x^* \in L^\infty$ and $x \in L^1$.

Definition 2. When a linear operator, T on L^1, satisfies the following conditions, we call T a *positive contraction*:

1. if $\|T\| \leq 1$;

2. for any positive function $f \geq 0$ a.e., $Tf \geq 0$ a.e.

Theorem 1. *Let T be a positive contraction on L^1. For any $A \in \mathscr{E}$, a map*

$$K(x, A) = T^* \mathbf{1}_A(x) \tag{1}$$

satisfies the following conditions, where T^* is the adjoint operator of T, and $\mathbf{1}_A$ is the indicator function of A:

1. for any fixed set $A \in \mathscr{E}$, the function $K(\cdot, A)$ is measurable;

2. for almost every $x \in E$, the set function $K(x, \cdot)$ is a measure on $(E, \mathscr{E})$;

3. for almost every $x \in E, K(x, E) \leq 1$.

Proof.

1. For any $A \in \mathscr{E}, T^*\mathbf{1}_A \in L^\infty(E, \mathscr{E}, m)$;

2. It is obvious that $K(x, \emptyset) = T^*\mathbf{1}_\emptyset = 0$ a.e. since T^* is a positive contraction on $L^\infty, K(x, A) = T^*\mathbf{1}_A(x) \geq 0$ a.e. for any $A \in \mathscr{E}$.

   Let $A = \cup_{i=1}^\infty A_i$ for pairwise disjoint $A_i \in \mathscr{E}$. For any $u \in L^1, u \geq 0$,

$$\begin{aligned}
\langle T^*\mathbf{1}_A, u\rangle &= \langle \mathbf{1}_A, Tu\rangle \\
&= \left\langle \mathbf{1}_{\cup_{i=1}^\infty A_i}, Tu \right\rangle = \int_{\cup_{i=1}^\infty A_i} Tu(x)m(dx) \\
&= \sum_{i=1}^\infty \int_{A_i} Tu(x)m(dx) = \sum_{i=1}^\infty \langle \mathbf{1}_{A_i}, Tu\rangle = \sum_{i=1}^\infty \langle T^*\mathbf{1}_{A_i}, u\rangle \\
&= \sum_{i=1}^\infty \int_E u(x) \cdot T^*\mathbf{1}_{A_i}(x)m(dx) = \lim_{n\to\infty} \int_E \sum_{i=1}^n u(x) \cdot T^*\mathbf{1}_{A_i}(x)m(dx) \\
&= \lim_{n\to\infty} \int_E u(x) \cdot \sum_{i=1}^n T^*\mathbf{1}_{A_i}(x)m(dx)
\end{aligned} \tag{2}$$

Since $u, \ T^*\mathbf{1}_{A_i} \geq 0, u \cdot \sum_{i=1}^n T^*\mathbf{1}_{A_i} \uparrow u \cdot \sum_{i=1}^\infty T^*\mathbf{1}_{A_i}$ a.e. as $n \to \infty$. We have

$$\lim_{n\to\infty} \int_E u(x) \cdot \sum_{i=1}^n T^*\mathbf{1}_{A_i}(x)m(dx) = \int_E u(x) \cdot \sum_{i=1}^\infty T^*\mathbf{1}_{A_i}(x)m(dx) = \int_E u(x) \cdot T^*\mathbf{1}_A(x)m(dx) \tag{3}$$

by the monotone convergence theorem. This means that

$$\langle T^*\mathbf{1}_A, u\rangle = \left\langle \sum_{i=1}^\infty T^*\mathbf{1}_{A_i}, u \right\rangle \tag{4}$$

for any $u \in L^1, u \geq 0$. Since every $u \in L^1$ is represented by $u = u^+ - u^-, u^+, u^- \in L^1, u^+, u^- \geq 0,$ we have

$$\begin{aligned}
\langle T^*\mathbf{1}_A, u\rangle &= \langle T^*\mathbf{1}_A, u^+\rangle - \langle T^*\mathbf{1}_A, u^-\rangle \\
&= \left\langle \sum_{i=1}^\infty T^*\mathbf{1}_{A_i}, u^+ \right\rangle - \left\langle \sum_{i=1}^\infty T^*\mathbf{1}_{A_i}, u^- \right\rangle \\
&= \left\langle \sum_{i=1}^\infty T^*\mathbf{1}_{A_i}, u \right\rangle.
\end{aligned} \tag{5}$$

Therefore, $K(x, \cdot)$ is completely additive, that is, if $A = \bigcup_{i=1}^\infty A_i$ for disjoint $A_i \in \mathscr{E}$, then

$$K(x,A) = \sum_{i=1}^{\infty} K(x,A_i) \qquad \text{for a.e. } x. \tag{6}$$

3. Since T^* is a positive contraction on $L^\infty, K(x,E) = T^*\mathbf{1}_E(x) \leq \mathbf{1}_E(x)$ a.e.

Remark 1. If the results 1–3 of theorem 1 hold for everywhere $x \in E$, then $K(x, A)$ is stochastic kernel (see Definition 1); see [7, 8] for more detail.

Secondly, we show that any stochastic kernel induces the positive contraction on L^1-space.

Theorem 2. *Let $K : E \times \mathcal{E} \to [0,1]$ be a stochastic kernel satisfying the condition*

$$K(x,A) = 0 \qquad a.e. \ x \in E \tag{7}$$

for any $A \in \mathcal{E}, m(A) = 0$. Then, for any absolutely continuous measure μ with respect to m, the measure ν on $\mathcal{E}$:

$$\nu(A) = \int_E K(x,A)\mu(dx) \qquad A \in \mathcal{E} \tag{8}$$

is an absolutely continuous measure with respect to m, and there exists the positive contraction T on L^1 such that

$$\tilde{\nu} = T\tilde{\mu}, \tag{9}$$

where $\tilde{\mu}, \tilde{\nu} \in L^1$ are the Radon-Nikodym derivatives of μ, ν, respectively.

Proof. We show that for any absolutely continuous measure μ with respect to m,

$$\nu(A) = \int_E K(x,A)\mu(dx) \tag{10}$$

is an absolutely continuous measure with respect to m.

From the definitions 1 and 2 of a stochastic kernel, we have $\nu(\emptyset) = 0$ and $\nu(A) \geq 0$ for any $A \in \mathcal{E}$.

Let $A = \cup_{i=1}^{\infty} A_i$ for pairwise disjoint $A_i \in \mathcal{E}$. Since $K(x, A_i) \geq 0$, it follows that

$$\begin{aligned} \nu(A) &= \int_E K(x,A)\mu(dx) = \int_E K(x, \cup_{i=1}^{\infty} A_i)\mu(dx) \\ &= \int_E \sum_{i=1}^{\infty} K(x,A_i)\mu(dx) = \sum_{i=1}^{\infty} \int_E K(x,A_i)\mu(dx) = \sum_{i=1}^{\infty} \nu(A_i) \end{aligned} \tag{11}$$

by the monotone convergence theorem: hence, ν is a measure on $\mathcal{E}$.

From the condition, $K(x,A) = 0$ for any $A \in \mathcal{E}, m(A) = 0,$; thus, we have

$$\nu(A) = \int_E K(x,A)\mu(dx) = \int_E 0\mu(dx) = 0, \tag{12}$$

which means that ν is an absolutely continuous measure with respect to m.

Since m is a σ-finite measure, by the Radon-Nikodym theorem, there are the Radon-Nikodym derivatives $\tilde{\mu}, \tilde{\nu} \in L^1, \tilde{\mu}, \tilde{\nu} \geq 0$ such as

$$\mu(A) = \int_A \tilde{\mu}(x) m(dx) \quad \text{and} \quad \nu(A) = \int_A \tilde{\nu}(x) m(dx) \tag{13}$$

for any $A \in \mathcal{E}$. We can define an operator T on the positive cone L^1_+ of L^1 such that

$$\tilde{\nu} = T\tilde{\mu}. \tag{14}$$

For notational convenience, we shall write $T\mu$ for ν.

For any $f \in L^1$, the completely additive set function

$$\varphi(A) = \int_A f(x) m(dx) \qquad A \in \mathcal{E} \tag{15}$$

is absolutely continuous with respect to m. Let $\varphi = \varphi_+ - \varphi_-$ be the Jordan decomposition of φ. Both φ_+ and φ_- are absolutely continuous with respect to m; then there exist the Radon-Nikodym derivatives $f_+, f_- \in L^1_+$ of φ_+, φ_-, respectively and $f(x) = f_+(x) - f_-(x)$ a.e. $x \in E$. By using them, we define $T\varphi$ as follows:

$$\begin{aligned} T\varphi &= \int_E K(x,A)\varphi(dx) = \int_E K(x,A) f(x) m(dx) \\ &= \int_E K(x,A)(f_+(x) - f_-(x)) m(dx) = \int_E K(x,A) f_+(x) m(dx) - \int_E K(x,A) f_-(x) m(dx) \\ &= \int_E K(x,A)\varphi_+(dx) - \int_E K(x,A)\varphi\_m(dx) = T\varphi_+ - T\varphi_-. \end{aligned} \tag{16}$$

$T\varphi$ is also absolutely continuous with respect to m. Thus, there exist the Radon-Nikodym derivatives $g \in L^1$ of $T\varphi$, and we define $Tf = g$. From the definition of T, T is a positive linear operator on L^1.

Finally, we prove that T is a contraction on L^1. For any $f \in L^1$, we obtain

$$\|f\| = \int_E f_+(x) m(dx) + \int_E f_-(x) m(dx) = \varphi_+(E) + \varphi_-(E). \tag{17}$$

In a similar way, we can show that

$$\|Tf\| = T\varphi_+(E) + T\varphi_-(E). \tag{18}$$

Then we have

$$\begin{aligned} T\varphi_+(E) &= \int_E K(x,A)\varphi_+(dx) \leq \int_E \mathbf{1}_E(x)\varphi_+(dx) = \varphi_+(E), \\ T\varphi_-(E) &= \int_E K(x,A)\varphi_-(dx) \leq \int_E \mathbf{1}_E(x)\varphi_-(dx) = \varphi_-(E). \end{aligned} \tag{19}$$

This implies that $\|Tf\| \leq \|f\|$.

We remark that the condition (7) in Theorem 2 is necessary.

Lemma 1 ([8]). *A stochastic kernel $K : E \times \mathcal{E} \to [0,1]$ satisfies the condition (7) if and only if for any absolutely continuous measure μ with respect to m,*

$$\nu(A) = \int_E K(x,A)\mu(dx) \tag{20}$$

is an absolutely continuous measure with respect to m.

Proof. If K does not satisfy (7), then there exist a set, $A \in \mathcal{E}$, and an element, $x \in E$, such that $m(A) = 0$, and $K(x,A) > 0$. We define a set, $B = \{x \in E : K(x,A) > 0\} \in \mathcal{E}$. B is not empty, and $m(B) > 0$. The measure μ on $\mathcal{E}$ defined by $\mu(C) = m(C \cap B)$ for any $C \in \mathcal{E}$ is absolutely continuous with respect to m, but $\nu(A) = \int_E K(x,A)\mu(dx) = \int_B K(x,A)m(dx) > 0$.

3. Markov operators induced by a dynamical system

In the previous section, we showed the relations between a stochastic kernel and a positive contraction. In this section, we shall treat a certain kind of positive contraction, which is called a Markov operator.

Firstly, we give the definition of a Markov operator on L^1-space (see [9–11]). Let E be a nonempty set and $\mathcal{E}$ a σ-algebra of subsets of E. Let $(E, \mathcal{E}, m)$ be a measure space, $L^1(E, \mathcal{E}, m)$ be the space of real valued integrable functions with respect to $(E, \mathcal{E}, m)$ and $\langle x^*, x\rangle$ be a dual pair of $x^* \in L^\infty$ and $x \in L^1$.

Definition 3. A liner operator T in an $L^1(E, \mathcal{E}, m)$ is called a *Markov operator* if

$$Tf \geq 0 \quad \text{and} \quad \|Tf\| = \|f\| \qquad \text{for all } f \in L^1_+(E, \mathcal{E}, m), \tag{21}$$

where $\|\cdot\|$ is the $L^1(E, \mathcal{E}, m)$-norm, and L^1_+ is the positive cone of $L^1(E, \mathcal{E}, m)$.

From this definition, obviously, Markov operators are positive operators and satisfy the following properties.

Preposition 1. *Markov operators are linear contraction in L^1-space.*

Proof. Since T is a positive operator, we have that for m-almost everywhere $x \in E$

$$\begin{aligned}
(Tf)^+(x) &= \left(Tf^+(x) - Tf^-(x)\right)^+ = \max\{0,\ Tf^+(x) - Tf^-(x)\} \\
&\leq \max\{0, Tf^+(x)\} = Tf^+(x); \\
(Tf)^-(x) &= \left(Tf^+(x) - Tf^-(x)\right)^- = \max\{0,\ Tf^-(x) - Tf^+(x)\} \\
&\leq \max\{0, Tf^-(x)\} = Tf^-(x),
\end{aligned} \tag{22}$$

where $f^+(x) = \max\{0, f(x)\}$ and $f^-(x) = \max\{0, -f(x)\}$. Thus, it follows that

$$|Tf| = (Tf)^+ + (Tf)^- \leq Tf^+ + Tf^- = T(f^+ + f^-) = T|f| \quad \text{a.e.} \tag{23}$$

From this inequality and the second condition of the definition of Markov operator, we have that

$$\|Tf\| = \int_E |Tf(x)| m(dx) \leq \int_E T|f(x)| m(dx) = \int_E |f| m(dx) = \|f\|. \tag{24}$$

This implies that T is a contraction in L^1-space.

Preposition 2. *The second condition of Markov operator: that is,* $\|Tf\| = \|f\|$ *for* $f \in L^1_+$*, is equivalently expressed as* $T^*\mathbf{1}_E = \mathbf{1}_E$*, where* T^* *is the adjoint operator of T, and* $\mathbf{1}_E(x)$ *is the indicator function of E.*

Proof. By the second condition of Markov operator, we have for any $C \in \mathscr{E}$

$$\int_C T^*\mathbf{1}_E(x)m(dx) = \langle T^*\mathbf{1}_E,\ \mathbf{1}_C\rangle = \langle \mathbf{1}_E,\ T\mathbf{1}_C\rangle = \langle \mathbf{1}_E,\ \mathbf{1}_C\rangle = \int_C \mathbf{1}_E(x)m(dx). \quad (25)$$

This implies that $T^*\mathbf{1}_E = \mathbf{1}_E$.

Conversely, if $T^*\mathbf{1}_E = \mathbf{1}_E$, then we have for $f \in L^1_+$

$$\|Tf\| = \langle \mathbf{1}_E,\ Tf||\rangle = \langle \mathbf{1}_E,\ Tf\rangle = \langle T^*\mathbf{1}_E,\ f\rangle = \langle \mathbf{1}_E,\ f\rangle = \|f\|. \quad (26)$$

Remark 2. Notice that sometimes a Markov operator is considered as a positive linear contraction defined on L^1-space because of Theorem 2, and there are many important results about positive linear contraction in the Ergodic theory. For example, the splitting theorem of Jacobs-Deleeuw-Glicksberg (see [8]) and "zero-two"law by Ornstein and Sucheston (see [12], 1970).

Next, we consider a measurable function, $K : E \times E \to \mathbb{R}$ such that

$$K(x,y) \geq 0 \qquad \text{for all } x, y \in E \quad (27)$$

and

$$\int_E K(x,y)m(dx) = 1 \qquad \text{for a.e. } y \in E. \quad (28)$$

By (27) and (28),

$$\begin{aligned}\int_E\left(\int_E |K(x,y)f(y)|m(dx)\right)m(dy) &= \int_E\left(\int_E K(x,y)m(dx)\right)|f(y)|m(dy)\\ &= \int_E |f(y)|m(dy) = \|f\|\end{aligned} \quad (29)$$

for any $f \in L^1$; hence, $|K(x,y)f(y)|$ is integrable on $E \times E$ and

$$\begin{aligned}\|f\| = \int_E\left(\int_E |K(x,y)f(y)|m(dx)\right)m(dy) &= \int_E\left(\int_E |K(x,y)f(y)|m(dy)\right)m(dx)\\ \geq \int_E\left|\int_E K(x,y)f(y)m(dy)\right|m(dx)\end{aligned} \quad (30)$$

by Fubini's Theorem (see). Thus, we can define an integral operator T on L^1 by

$$Tf(x) = \int_E f(y)K(x,y)m(dy). \quad (31)$$

We have

$$\|f\| \geq \|Tf\|. \quad (32)$$

Then, T is a positive contraction on L^1. Furthermore, T is a Markov operator on L^1. Indeed, we have that for any $f \in L^1_+$,

$$\begin{aligned}\|Tf\| &= \int_E Tf(x)m(dx) = \int_E \left(\int_E f(y)K(x,y)m(dy)\right)m(dx) \\ &= \int_E \left(\int_E K(x,y)m(dx)\right)f(y)m(dy) = \int_E f(y)m(dy) \\ &= \|f\|.\end{aligned} \tag{33}$$

For the sake of convenience and without loss of generality, we shall assume that E is $\mathbb{R}$, $\mathscr{E}$ the σ-algebra of Borel subsets, and λ the Lebesgue measure on $\mathbb{R}$. We shall give some stochastic processes, which induce Markov operators as an integral operator defined by (31) for a given measurable function $K(x,y)$.

Let $(\Omega, \mathscr{F}, P)$ be a probability space, where $\mathscr{F}$ denotes a Borel σ-field and P a probability measure, $X_0, Y_0, Y_1, \cdots$ independent $\mathbb{R}$-valued random variables with densities f_0 and $g_n = g$, respectively, and $S : \mathbb{R} \to \mathbb{R}$ be a measurable transformation. Suppose that $g \in D := \{f \in L^1 : \ \|f\| = 1, f \geq 0\}$. Under these conditions, we shall give examples of two types of random dynamical systems, for which Markov operators are induced.

- **Additive noise type:** Consider the following stochastic process $\{X_n\}$ defined by

$$X_{n+1} = S(X_n) + Y_n \tag{34}$$

for each $n \geq 0$.

Since X_0 and $\{Y_n\}$ are independent and f_n, g are integrable, we have that

$$\begin{aligned}P(X_{n+1} \in A) = P(S(X_n) + Y_n \in A) &= \iint_{S(x)+y\in A} f_n(y)g(x)\lambda(dy)\lambda(dx) \\ &= \int_{z\in A}\left\{\int_{y\in\mathbb{R}} f_n(y)g(z - S(y))\lambda(dy)\right\}\lambda(dz) \\ &= \int_A T^n f_0(z)\lambda(dz).\end{aligned} \tag{35}$$

If we set $K(x,y) = g(x - S(y))$, then $K(x,y)$ satisfies (27) and (28). Thus, the density function f_n of X_n is represented by the n-th iteration of the linear operator $T : L^1(\mathbb{R}) \to L^1(\mathbb{R})$. Now we show that T is a Markov operator. Since $g \geq 0$ and $\|g\| = 1$, we have that for any $f \in L^1(\mathbb{R})$ with $f \geq 0$

$$\begin{aligned}\|Tf\| &= \int_{\mathbb{R}}\left\{\int_{\mathbb{R}} f(y)g(x - S(y))\lambda(dy)\right\}\lambda(dx) \\ &= \int_{\mathbb{R}} f(y)\left\{\int_{\mathbb{R}} g(x - S(y))\lambda(dx)\right\}\lambda(dy) \\ &= \int_{\mathbb{R}} f(y)\lambda(dy) = \|f\|\end{aligned} \tag{36}$$

by the Fubini's theorem. Therefore, $T : L^1(\mathbb{R}) \to L^1(\mathbb{R})$ is a Markov operator.

- **Example** [13]: Let $X_0, Y_0, Y_1, \cdots$ be independent $\mathbb{R}$-valued random variables with densities f_0 and $g_n = g \in D$, respectively, and $S : \mathbb{R} \to \mathbb{R}$ be a Borel measurable transformation, satisfying

$$|S(x)| \leq \alpha|x| + \beta \quad \text{for } x \in \mathbb{R}, \tag{37}$$

where $0 < \alpha < 1$ and $\beta > 0$.
We assume that the density g of Y_n has a finite first moment, that is,

$$\int_{\mathbb{R}} |x| g(x) \lambda(dx) < \infty. \tag{38}$$

Under these conditions, we consider the following stochastic process $\{X_n\}$, that is, $\{X_n\}$ is the dynamical system perturbed by additive noises:

$$X_{n+1} = S(X_n) + Y_n \qquad \text{for} \qquad n = 0, 1, \dots . \tag{39}$$

In this case, the Markov operator $T : L^1(\mathbb{R}) \to L^1(\mathbb{R})$ defined by

$$Tf(x) = \int_{\mathbb{R}} f(y) g(x - S(y)) \lambda(dy) \tag{40}$$

is *weakly constrictive*, that is, there exists the weakly precompact set $\mathscr{F} \subset L^1(\mathbb{R})$ such that

$$\lim_{n\to\infty} d(T^n f, \mathscr{F}) = 0 \tag{41}$$

for $f \in D$, where $d(f, \mathscr{F})$ denotes the distance with respect to the norm $\| \cdot \|$ between f and the set $\mathscr{F}$. This implies that $L^1(\mathbb{R}) = E_{fl}(T) \oplus E_{rev}(T)$ with

$$E_{fl}(T) = \left\{ x \in X : \lim_{n\to\infty} \|T^n x\| = 0 \right\} \quad \text{and} \quad \dim(E_{rev}(T)) < \infty. \tag{42}$$

Furthermore, Lasota and Mackey show numerically that the limiting densities of $\{X_n\}$ become asymptotically periodic with period 3 [13], and refer to Section 3 for more details.

- **Multiplicative noise type:** Consider the following stochastic process $\{X_n^\varepsilon\}_{n\geq 0}$ defined by

$$X_{n+1}^\varepsilon(\omega) = (1 - \varepsilon Y_n) S(X_n^\varepsilon(\omega)) \qquad \text{for all } n \geq 0, \tag{43}$$

where $X_0^\varepsilon = X_0$ for each $0 < \varepsilon < 1$.
By a similar argument as additive noise, the density function f_n^ε of X_n^ε is represented by the n-th iteration of the linear operator T_ε because X_0 and $\{Y_n\}$ are independent. Actually, if we assume that there exists the density function f_n^ε of X_n^ε, then we have the following formula for any Borel subset $A \subset \mathbb{R}$:

$$\begin{aligned} P(X_{n+1}^\varepsilon \in A) &= \iint_{xS(y) \in A} f_n^\varepsilon(y) \frac{1}{\varepsilon} g\left(\frac{1-x}{\varepsilon}\right) \lambda(dy)\lambda(dx) \\ &= \int_A \left\{ \int_{\mathbb{R}} f_n^\varepsilon(y) g\left(\frac{1}{\varepsilon}\left(1 - \frac{x}{S(y)}\right)\right) \frac{1}{\varepsilon S(y)} \lambda(dy) \right\} \lambda(dx) \\ &= \int_A T_\varepsilon f_n^\varepsilon(x) \lambda(dx). \end{aligned} \tag{44}$$

This implies that $P(X^{\varepsilon}_{n+1} \in A) = \int_A T^{n+1}_{\varepsilon} f_0(x)\lambda(dx)$. Moreover, it is easy to see that T_ε is the Markov operator, that is, $T_\varepsilon f \geq 0$ and $\|T_\varepsilon f\| = \|f\|$ for any $f \in L^1([0,1])$ with $f \geq 0$ because

$$\begin{aligned} \|T_\varepsilon f\| &= \int_{[0,1]} f(y) \left\{ \int_{\left[\frac{1}{\varepsilon} - \frac{1}{\varepsilon S(y)}, \frac{1}{\varepsilon}\right] \cap [0,1]} g(x)\lambda(dx) \right\} \lambda(dy) = \int_{[0,1]} f(y)\lambda(dy) \\ &= \|f\| \end{aligned} \tag{45}$$

by Fubini's theorem (see [14] for more details). In this case, we can study how the stochastic process $\{X^{\varepsilon}_n\}$ is changed as $\varepsilon \to 0$.

- **Example** [14]: Let $S : [0,1] \to [0,1]$ be a non-singular transformation and $\{Y_n\}_{n \geq 0}$ be an i.i.d. sequence. We assume that the density function g for $\{Y_n\}_{n \geq 0}$ has full support in $[0, 1]$, and the following conditions are satisfied:

  i. There exists a partition $0 = a_1 < a_2 < \cdots < a_m = 1$ such that for each integer $j = 1, \cdots, m-1$, the restriction S_j of S to the interval $[a_j, a_{j+1})$ is a monotone C^1-function and

$$c_j := \inf_{x \in [a_j, a_{j+1})} \left\{ |S_j'(x)| \right\} > 0, \quad c_{m-1} := \inf_{x \in [a_{m-1}, a_m]} \left\{ |S_j'(x)| \right\} > 0 \tag{46}$$

  for $j = 1, \cdots, m-2$.

  ii. $g \in L^\infty(\mathbb{R})$;

  iii. $c_1 > \frac{1}{1-\varepsilon_0}$ for some $\varepsilon_0 \in (0,1)$.

Under these conditions, we consider the stochastic process $\{X^{\varepsilon}_n\}$ defined by (43), where $\varepsilon < \varepsilon_0$. The density functions of $\{X^{\varepsilon}_n\}$ are represented by the Markov operator

$$T_\varepsilon f(x) = \int_{[0,1] \setminus \{0\}} P_S f(y) g\left(\frac{1}{\varepsilon}\left(1 - \frac{x}{y}\right)\right) \frac{1}{\varepsilon y} \lambda(dy). \tag{47}$$

In this case, the Markov operator T_ε is weakly constrictive for any $\varepsilon < \varepsilon_0$. Moreover, if $g(x) > 0$ for all $x \in [0,1]$, then the Markov operators T_ε is asymptotically stable; that is, there is a unique $f^{\varepsilon}_{*} \in D$ such that $T_\varepsilon f^{\varepsilon}_{*} = f^{\varepsilon}_{*}$ and

$$\lim_{n \to \infty} \|T^n_\varepsilon f - f^{\varepsilon}_{*}\|_1 = 0 \qquad \text{for all } f \in D, \tag{48}$$

where $D = \{f \in L^1(\mathbb{R}) \mid f \geq 0, \|f\| = 1\}$.

Remark 3. If $S : \mathbb{R} \to \mathbb{R}$ is a non-singular measurable transformation (i.e., $\lambda(S^{-1}(A)) = 0$ for any Borel set $A \subset \mathbb{R}$ with $\lambda(A) = 0$), then there exists a linear operator, $T_S : L^1(\mathbb{R}) \to L^1(\mathbb{R})$, defined by

$$\int_A T_S f_0(x)\lambda(dx) = \int_{S^{-1}(A)} f_0(x)\lambda(dx) \qquad \text{for Borel set } A \subset \mathbb{R}, \tag{49}$$

and called the *Perron-Frobenius operator* corresponding to S.

4. The iteration of a linear contractive operator on Banach space

The iteration of a positive linear contraction expresses the asymptotic behavior of a Markov process. Let E be a real or complex Banach space and T be a contractive linear operator on E. If E can be decomposed into the direct sum

$$E = E_{fl}(T) \oplus E_{rev}(T) \tag{50}$$

with respect to T, where

$$\begin{aligned} E_{fl}(T) &:= \{x \in E: \ 0 \in \text{w-cl}\{T^n x, \ n \in \mathbb{N} \cup \{0\}\}\}, \quad \text{and} \\ E_{rev}(T) &:= \{x \in E: \ y \in \text{w-cl}\{T^n x, \ n \in \mathbb{N} \cup \{0\}\} \ \Rightarrow \ x \in \text{w-cl}\{T^n y, \ n \in \mathbb{N} \cup \{0\}\}\}, \end{aligned} \tag{51}$$

then we call this decomposition the Jacobs-de Leeuw-Glicksberg decomposition (see [8, 10, 15]). This decomposition plays a very important role in this section.

Definition 4 [16]. We call a linear contraction T on a Banach space $(E, \|\cdot\|)$ *constrictive* if there is a compact subset, $A \subset E$ such that

$$\lim_{n\to\infty} \inf_{y \in A} \|T^n x - y\| = 0 \qquad \text{for each } x \in B(E), \tag{52}$$

where $B(E)$ is the closed unit ball of E. We call A a *constrictor* for T.

Theorem 3 [17]. *Let E be a strictly convex and reflexive complex Banach space with the strictly convex dual space E^*. Given a linear contractive operator T on a complex Banach space E, the following assertions are equivalent:*

1. T is constrictive.

2. $x \in E_{fl}(T) = \{x \in X: \ \lim_{n\to\infty} \|T^n x\| = 0\}$ if and only if $\langle x, h^* \rangle = 0$ holds for all eigenvectors h^* of T^* having unimodular eigenvalues.

Theorem 4 [17]. *Let (Ω, Σ, μ) be a finite measure space and T be a Markov operator on real $L^1(\Omega)$ defined by (31), where $K : \Omega \times \Omega \to \mathbb{R}$ is a measurable function that satisfies (27) and (28). Consider the following assertions:*

1. *T satisfies $\|Tx\|_p \le \|x\|_p$ for some $1 \le p < \infty$, where $\|\cdot\|_p$ denotes the L^p-norm.*

2. *The sub σ-algebra*

$$\Sigma_0(T) = \{A \in \Sigma: \ T^n \mathbf{1}_A = \ \textit{characteristic function} \ \ \forall n \ge 0\} \tag{53}$$

has at most finitely many atoms.

3. *For each atom $W \in \Sigma_0(T)$,*

$$\lim_{n\to\infty} \mu(A \backslash \text{supp}(T^{dn} \mathbf{1}_B)) = 0, \tag{54}$$

for all subset $A, B \subset W$ with $\mu(A), \mu(B) > 0$, where d is the least common multiple of orders of atoms in $\Sigma_0(T)$.

Then T is constrictive on $L^p(\Omega)$ and

$$\begin{aligned} &L^p(\Omega) = E_{rev}(T) \oplus E_{fl}(T), \\ &E_{rev}(T) = \overline{\text{spn}}\{\mathbf{1}_W : \ W \in \Sigma_0(T) \ \ is \ \ atom \ \ \}, and \\ &E_{fl}(T) = \left\{ x \in L^p(\Omega) : \ \lim_{n\to\infty} \|T^n x\|_p = 0 \right\}. \end{aligned} \tag{55}$$

Author details

Takashi Honda[1*] and Yukiko Iwata[2]

1 Faculty of Education, Department of Mathematics, Iwate University, Morioka, Japan

2 Faculty of Liberal Arts, Tohoku Gakuin University, Sendai, Japan

*Address all correspondence to: thonda7@iwate-u.ac.jp

References

[1] Honda T. A nonlinear analytic approach to the splitting theorem of Jacobs-de Leeuw-Glicksberg. Nonlinear Analysis. 2012;**75**(16):6002-6008

[2] Iwata Y. Constrictive Markov operators induced by Markov processes. Positivity. 2016;**20**(2):355-367

[3] Jacobs K. Noise and statistical periodicity. Physica D: Nonlinear Phenomena. 1987;**28**:143-154

[4] Meyn S, Tweedie RL. Markov chains and stochastic stability. In: With a prologue by Peter W. Glynn. Second ed. Cambridge, New York, Melbourne, Madrid, Cape Town, Singapore, São Paulo, Delhi: Cambridge University Press; 2009

[5] Nummelin E. General irreducible Markov chains and nonnegative operators. In: Cambridge Tracts in Mathematics, 83. Cambridge: Cambridge University Press; 1984

[6] Revuz D. Markov chains. In: North-Holland Mathematical Library. Second ed. Vol. 11. Amsterdam: North-Holland Publishing Co.; 1984

[7] Foguel SR. The ergodic theory of Markov processes. In: Van Nostrand Mathematical Studies. Vol. 21. New York-Toronto, Ont.-London: Van Nostrand Reinhold Co; 1969

[8] Krengel U. Ergodic Theorems. Berlin: Walter de Gruyter & Co.; 1985

[9] Eisner T, Bálint F, Haase M, Nagel R. Operator theoretic aspects of ergodic theory. In: Graduate Texts in Mathematics. Vol. 272. Cham, Heidelberg, New York, Dordrecht, London: Springer; 2015

[10] Emel'yanov E, Eduard Y. Non-spectral asymptotic analysis of one-parameter operator semigroups. In: Operator Theory: Advances and Applications. Vol. 173. Basel: Birkhäuser Verlag; 2007

[11] Lasota A, Mackey MC. Chaos, fractals, and noise. Stochastic Aspects of Dynamics. In: Applied Mathematical Sciences, 97. Second ed. New York: Springer-Verlag; 1994

[12] Ornstein D, Sucheston L. An operator theorem on L^1 convergence to zero with applications to Markov kernels. Annals of Mathematical Statistics. 1970;**41**:1631-1639

[13] Lasota A, Mackey MC. Noise and statistical periodicity. Physica. 1987;**28D**: 143-154

[14] Iwata Y. Multiplicative stochastic perturbations of one-dimensional maps. Stochastics and Dynamics. 2013;**13**(2): 1-10

[15] de Leeuw K, Glicksberg I. Applications of almost periodic compactifications. Acta Mathematica. 1961;**105**:63-97

[16] Sine R. Constricted systems. The Rocky Mountain Journal of Mathematics. 1991;**21**(4):1373-1383

[17] Honda T, Iwata Y. Operator theoretic phenomena of the Markov operators which are induced by stochastic difference equations. In: Advances in Difference Equations and Discrete Dynamical Systems, Springer Proc. Math. Stat. Vol. 212. Singapore: Springer; 2017. pp. 125-135

Chapter 2

Performance Modeling of Finite Sources Retrial Queue Using Markov Regenerative Approach

Lyes Ikhlef

Abstract

We study a finite sources retrial queue with deterministic service times using an approach based on the theory of Markov regenerative process. A Deterministic Stochastic Petri Net (*DSPN*) model which copes with the complexity of this queue is given. For the steady state of this model, we construct the one-step transition probability matrix of Embedded Markov Chain (EMC) and the conversion matrix. We establish an algorithm based on the theoretic results obtained in order to compute efficiently various performance measures and to study the effect of system parameter's on the characteristics of the $M/D/1/N/N$ retrial queue.

Keywords: Markov regenerative process, stochastic Petri nets, embedded Markov chain, steady state, retrial queue

1. Introduction

The performance analysis of queueing systems has been studied over the past years. There are many models that have been developed in the queueing systems analysis using stochastic process, particularly Markov process, semi-Markov process, and Markov regenerative process. Different approaches have been explored in the literature dealing with non-Markovian models: method of supplementary variable [1], approximate analysis by phase type expansion [2], Markov renewal theory [3], etc.

The standard queueing systems with deterministic service times are widely found in the literature Bunday [4], Kashyap and Chaudhry [5], Bhat [6]. Brun and Garcia [7] give an analytical solution of the system $M/D/1/K$. Franx et al. [8] study the multi-server system $M/D/c$. Madan and Saleh [9] study the queue $M/D/1$ with general vacations. Choi et al. in [3, 10] investigate the transient and sensitivity analysis of *DSPN*. As an application, they detailed the analysis of the classical queue $M/D/1/2/2$ and $M/D/1/2/2$ with vacation. However, little attention has been paid to retrial queues with deterministic service times and finite sources [11]. Wu and Ke in [12], consider an infinite single server retrial queueing system in which each customer (primary or secondary customer) has discrete service times. For bibliographies on retrial queues, see [13–16] and the references therein.

In this paper, we give a model and performance analysis of the finite sources retrial system with deterministic service times by using *DSPN* tool.

2. Analysis of *DSPN* class

The *DSPN* class was introduced by Ajmone and Chiola [17]. The *DSPN* class was proposed to study a qualitative and quantitative behavior of non-Markovian systems. This class allows transitions with zero firing times (immediate transitions), exponentially distributed or deterministic distributed firing times. A *DSPN* describes a stochastic process referred to as the marking process $M(t), t \geq 0$. This marking process can not mapped into a Continuous Time Markov Chain (*CTMC*) [18]. Choi et al. have shown that, under the restriction the deterministic transitions are mutually exclusive, the underlying stochastic process of the DSPN class is a Markov regenerative process [3]. This process has a sequence of embedded time points $T_0, T_1, \ldots$ such that the states $Y_0, Y_1, \ldots$ of the process at these time points satisfy the Markov property. These time points are indicated by the Markov regeneration instants (regenerative time points, *RTP*). Note that the stochastic proces $(Y_n)_{n \geq 0}$ is an *EMC*. The steady-state solution of the *DSPN*, based on the Markov renewal method, can be summarized in the following steps:

- Constructing the two kernels: global kernel $K(t) = [K_{ij}(t)]$ and local kernel $E(t) = [E_{ij}(t)]$, where:

  a. $K_{ij}(t) = P(Y_1 = j, T_1 \leq t | Y_0 = i), i, j \in \Omega$ is the probability that the system transits from state i to state j upon firing of a general transition, or upon the firing of an exponential transition which is computed with a general transition.

  b. $E_{ij}(t) = P(M(t) = j, T_1 > t | Y_0 = i), i, j \in \Omega' \subset \Omega$ is the probability that the system transits from state i to state j while the general transition and the exponential that is competing with a general transition do not firing. The matrix $E(t)$ describes the behavior of the marking process between two regeneration instants.

- Obtaining the one step transition probability matrix $P = [P_{ij}] = K(\infty)$ of the *EMC* defined at *RTP*, $\{T_n, n \geq 1\}$, and the matrix $\alpha = [\alpha_{ij}]$. Here, the elements α_{ij} represent the mean time that the *DSPN* spends in state j between two successive regeneration instants, given that it started in state i after the last regeneration. They are given by:

$$\alpha_{ij} = \int_0^{\infty} E_{ij}(t)dt. \quad (1)$$

- Computing υ the steady state probability vector of the *EMC* by solving the linear system:

$$\begin{cases} vP & = v, \\ ve & = 1. \end{cases} \tag{2}$$

Where e is a column vector of ones.

- Computing the steady state probability distribution vector π of the *DSPN*, by the formula:

$$\pi_j = \frac{\sum\limits_{k \in \Omega} v_k \alpha_{kj}}{v_k u_k} = \sum_{k \in \Omega} \beta_k \frac{\alpha_{kj}}{u_k}, \tag{3}$$

where,

$$\mu_i = \sum_{j \in \Omega} \alpha_{ij} \text{ and } \beta_k = \frac{v_k u_k}{\sum\limits_{k \in \Omega} v_r u_r}. \tag{4}$$

We define the conversion matrix $C = [C_{ij}]$ by,

$$C_{ij} = \frac{\alpha_{ij}}{u_i}. \tag{5}$$

Let a vector $\beta = (\beta_k)$, the Eq. (3) can be rewritten in the matrix form:

$$\pi = \beta C. \tag{6}$$

3. Model description

We consider the retrial queue, $M/D/1/N/N$, in which primary customers arrive according to a Poisson process with rate λ. If an arriving customer finds the server idle, he obtains service immediately and joins the sources after service completion. Otherwise, if the server is occupied, the arriving primary customer enters into the orbit. The policy of access from the orbit to the server is governed by an exponential law with rate $k\gamma$, where k is the number of customers in the orbit. Each customer has a deterministic (constant) service times of length $\tau > 0$. The **Figure 1** shows the *DSPN* model describing the $M/D/1/N/N$ retrial queue.

This *DSPN* has four places *p.sour*, *p.chec*, *p.serv*, *p.orbi* noted by circles, two immediate transitions *t.acc*1, *t.acc*2 noted by thin black bars, two exponential (*EXP*) transitions *t.arri*, *t.retr* indicated by white rectangular boxes, and one deterministic (*DET*) transition *t.serv* noted by thick black rectangular box.

The vectors $(M(p.sour), M(p.chec), M(p.serv), M(p.orbi))$, describe the possible markings of the given *DSPN* model (i.e., where $M(p)$ represents the number of tokens in the place p). The vector $M_0 = (N,0,0,0)$ is its initial marking. The transitions are interpreted as follows:

- The *EXP* transition *t.arri* represents the arrival of the primary customers. The firing of this transition consists to remove a token in the place *p.sour* and to deposit a token in the place *p.chec*.

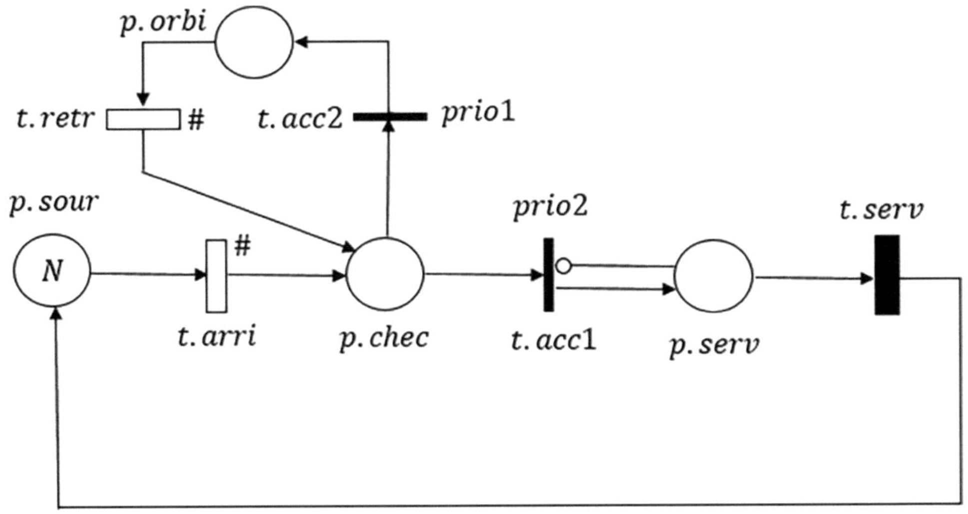

Figure 1.
DSPN models retrial queue M/D/1/N/N with classical retrial policy.

- The immediate transition $t.acc1$ represents the access to the service. The instantly firing of this transition consists to remove a token in the place $p.chec$ and to deposit a token in the place $p.serv$.

- The *DET* transition $t.serv$ represents the service of the customer. The firing of this transition consists to remove a token in the place $p.serv$ and to construct a token in the place $p.sour$.

- The transition $t.acc2$ represents the access to the orbit. The firing of this transition consists to destroy a token in the place $p.chec$ and to deposit a token in the place $p.orbi$.

- The *EXP* transition $t.retr$ represents the retrial customers. The firing of this transition consists to destroy a token in the place $p.orbi$ and to construct a token in the place $p.chec$. The firing rate of $t.arri$ and $t.retr$ are marking dependent and equals, respectively $M(p.sour)\lambda$, $M(p.orbi)\gamma$.

- The immediate transition $t.acc1$ has higher priority than the immediate transition $t.acc2$.

The interpretation of the places are given in the following **Table 1**.

Place name	Description	Initial marking
$p.sour$	Source of primary customers	N
$p.chec$	Contains primary customers or secondary customers	0
$p.serv$	Customers requiring a service	0
$p.orbi$	Orbit for the retrial customers	0

Table 1.
Interpretation of the places in the DSPN Model given in ***Figure 1****.*

4. EMC and steady-state results

In this section, we details the steady-state analysis of the *DSPN* models the $M/D/1/N/N$ retrial queue. The state transition diagram of the *DSPN* depicted in **Figure 1** is given in **Figure 2** with states space Ω such that:

$$\Omega = \{M_{2k} = (k, 0), M_{2k+1} = (k, 1), 0 \le k \le N-1\}. \tag{7}$$

Let $M_i, M_j \in \Omega$, the infinitesimal generator matrix $Q = \left[q_{M_iM_j}\right]$ of the subordinated *CTMC* with respect to *DET* transition *t.serv* is given by:

$$q_{M_iM_j} = \begin{cases} -\left(N - \dfrac{i+1}{2}\right)\lambda, & \text{if } \ 0 \le k \le N-2, i = 2k+1 \text{ and } j = i; \\ \left(N - \dfrac{i+1}{2}\right)\lambda, & \text{if } \ 0 \le k \le N-2, i = 2k+1 \text{ and } j = i+2; \\ 0, & \text{otherwise.} \end{cases} \tag{8}$$

The one step transition probability matrix $P = \left[P_{M_iM_j}\right]$ is given by:

$$P_{M_iM_j} = \begin{cases} 1, \ \text{if } \ i = 0 \ \text{ and } \ j = 1; \\ C_{N-\frac{i+1}{2}}^{\frac{j-i+1}{2}} \left(1 - e^{-\lambda\tau}\right)^{\frac{j-i+1}{2}} \left(e^{-\lambda\tau}\right)^{N-\frac{j+2}{2}}, \\ \text{if } 0 \le k \le N-1, i = 2k+1 \text{ and } i-1 \le j \le 2N-2, j = 2k; \\ \dfrac{\frac{i}{2}\gamma}{\frac{i}{2}\gamma + \left(N - \frac{i}{2}\right)\lambda}, \\ \text{if } 1 \le k \le N-1, i = 2k+1 \text{ and } j = i-1; \\ \dfrac{\left(N - \frac{i}{2}\right)\lambda}{\frac{i}{2}\gamma + \left(N - \frac{i}{2}\right)\lambda}, \\ \text{if } 1 \le k \le N-1, i = 2k+1 \text{ and } j = i+1; \\ 0, \ \ \text{otherwise.} \end{cases} \tag{9}$$

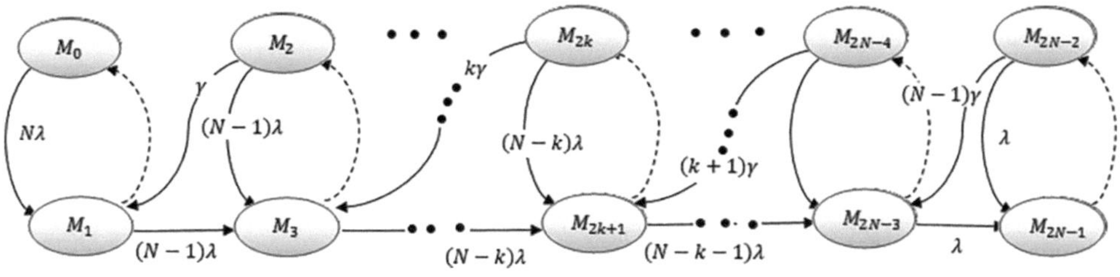

Figure 2.
Subordinated CTMC for the DSPN associated to M/D/1/N/N with classical retrial policy.

The conversion matrix $C = \left[c_{M_iM_j}\right]$ is given by:

$$c_{M_iM_j} = \begin{cases} 1, & \text{if } i = j = 0; \\ \frac{1}{\tau}\int_0^\tau C_{N-\frac{i+1}{2}}^{\frac{j-i}{2}} \left(1 - e^{-\lambda t}\right)^{\frac{j-i}{2}} \left(e^{-\lambda t}\right)^{N-\frac{j+1}{2}} dt, \\ \text{if } 0 \le k < N-1, i = 2k+1 \text{ and } i \le j \le 2N-1, j = 2k+1; \\ 1, & \text{if } 1 \le k \le N-1, i = 2k, \text{and } j = i; \\ 0, & \text{otherwise.} \end{cases} \tag{10}$$

5. Example

When $N = 2$, we obtain the state transition diagram (**Figure 3**) of the *DSPN* model depicted in **Figure 1**, with state space:

$$\Omega = \{M_0 = (0,0), M_1 = (1,0), M_2 = (0,1), M_3 = (1,1)\}. \tag{11}$$

The infinitesimal generator matrix $Q = \left[q_{M_iM_j}\right]$ of the subordinated *CTMC* with respect to *DET* transition, *t.serv*, is given by:

$$Q = \begin{pmatrix} 0 & 0 & 0 & 0 \\ 0 & -\lambda & 0 & \lambda \\ 0 & 0 & 0 & 0 \\ 0 & 0 & 0 & 0 \end{pmatrix}. \tag{12}$$

The one step transition probability matrix, $P = \left[P_{M_iM_j}\right]$, is given by:

$$P = \begin{pmatrix} 0 & 1 & 0 & 0 \\ e^{-\lambda\tau} & 0 & 1 - e^{-\lambda\tau} & 0 \\ 0 & \frac{\gamma}{\gamma+\lambda} & 0 & \frac{\lambda}{\gamma+\lambda} \\ 0 & 0 & 1 & 0 \end{pmatrix}. \tag{13}$$

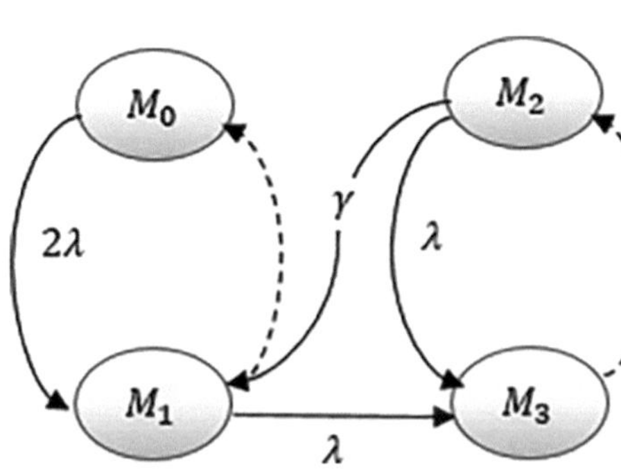

Figure 3.
Subordinated CTMC for the DSPN models M/D/1/2/2.

The steady state probabilities of the *EMC* are given by:

$$\begin{aligned} v_{M_0} &= \frac{1}{2}\frac{\gamma e^{-\lambda\tau}}{\gamma+\lambda-\lambda e^{-\lambda\tau}}, v_{M_1} = \frac{1}{2}\frac{\gamma}{\gamma+\lambda-\lambda e^{-\lambda\tau}}, \\ v_{M_2} &= \frac{1}{2}\frac{1-\left(e^{-\lambda\tau}\right)(\gamma+\lambda)}{\gamma+\lambda-\lambda e^{-\lambda\tau}}, v_{M_3} = \frac{1}{2}\frac{\lambda\left(1-e^{-\lambda\tau}\right)}{\gamma+\lambda-\lambda e^{-\lambda\tau}}. \end{aligned} \tag{14}$$

The elements $c_{M_iM_j}$ are given by the conversion matrix C:

$$C = \begin{pmatrix} 1 & 0 & 0 & 0 \\ 0 & \frac{1}{\lambda\tau}\left(1-e^{-\lambda\tau}\right) & 0 & 1+\frac{1}{\lambda\tau}\left(e^{-\lambda\tau}-1\right) \\ 0 & 0 & 1 & 0 \\ 0 & 0 & 0 & 1 \end{pmatrix}. \tag{15}$$

The steady state probabilities of the *DSPN* are given by:

$$\begin{aligned} \pi_{M_0} &= \frac{\gamma e^{-\lambda\tau}}{2\gamma\tau\lambda+2\lambda+2\lambda^2\tau+e^{-\lambda\tau}\left(\gamma-2\lambda-2\lambda^2\tau\right)}; \\ \pi_{M_1} &= \frac{2\gamma\left(1-e^{-\lambda\tau}\right)}{2\gamma\tau\lambda+2\lambda+2\lambda^2\tau+e^{-\lambda\tau}\left(\gamma-2\lambda-2\lambda^2\tau\right)}; \\ \pi_{M_2} &= \frac{2\lambda\left(1-e^{-\lambda\tau}\right)}{2\gamma\tau\lambda+2\lambda+2\lambda^2\tau+e^{-\lambda\tau}\left(\gamma-2\lambda-2\lambda^2\tau\right)}; \\ \pi_{M_3} &= \frac{2\left(-\gamma+\gamma\tau\lambda+\gamma e^{-\lambda\tau}+\lambda^2\tau-\lambda^2\tau e^{-\lambda\tau}\right)}{2\gamma\tau\lambda+2\lambda+2\lambda^2\tau+e^{-\lambda\tau}\left(\gamma-2\lambda-2\lambda^2\tau\right)}. \end{aligned} \tag{16}$$

The steady state probabilities v and π for λ = 0.5, γ = 0.2, τ = 1.5 are given in the following **Table 2**.

k	$v_{M_{2k}}$	$v_{M_{2k+1}}$	$\pi_{M_{2k}}$	$\pi_{M_{2k+1}}$
0	0.1018434	0.2156024	0.0716884	0.1601520
1	0.3981566	0.2843976	0.4003710	0.3677796

Table 2.
The steady state probabilities distributions of the DSPN models the $M/D/1/2/2$ retrial queue, "$\lambda = 0.5$, $\gamma = 0.2$, $\tau = 1.5$."

6. Performance measures

By using the steady-state probabilities distributions

$$\pi = \left(\pi_{M_0}, \pi_{M_1}, \ldots, \pi_{M_{2k}}, \pi_{M_{2k+1}}, \ldots, \pi_{M_{2N-2}}, \pi_{M_{2N-1}}\right) \tag{17}$$

the main system performance measures can be derived by the following formulas:

- The mean number of customers in the orbit (n_o):

$$n_o = \sum_{k=1}^{N-1} k \left[\pi_{M_{2k}} + \pi_{M_{2k+1}}\right]. \quad (18)$$

- The mean generation rate of primary calls ($\overline{\lambda}$):

$$\overline{\lambda} = \lambda \left[\sum_{k=0}^{N-1} (N-k)\, \pi_{M_{2k}} + \sum_{k=0}^{N-2} (N-k-1)\, \pi_{M_{2k+1}}\right]. \quad (19)$$

- The mean generation rate of repeated calls ($\overline{\gamma}$):

$$\overline{\gamma} = \gamma \sum_{k=1}^{N-1} k \left[\pi_{M_{2k}} + \pi_{M_{2k+1}}\right]. \quad (20)$$

- The mean generation rate of server ($\overline{\tau}$):

$$\overline{\tau} = \tau \sum_{k=0}^{N-1} \pi_{M_{2k+1}}. \quad (21)$$

- The server utilization (U_s):

$$U_s = \sum_{k=0}^{N-1} \pi_{M_{2k+1}}. \quad (22)$$

- The mean waiting time ($\overline{w}$) (Little's law):

$$\overline{w} = \frac{n_o}{\overline{\lambda}}. \quad (23)$$

- The mean response time (ϖ) (Little's law):

$$\varpi = \frac{n_s}{\overline{\lambda}}. \quad (24)$$

7. Numerical results

In this section, we give some numerical results, concerning the *DSPN* associated to $M/D/1/N/N$ retrial queue. The results obtained based on the following algorithm.

Algorithm
Begin
Step 0: Give N the number of sources, λ customer arrival rate,

τ deterministic service times, γ the retrial rate.
Step 1: Compute the one step transition probability matrix P and the conversion matrix C using
the Eqs. (9) and (10).
Step 2: Compute the stationary distribution v of the *EMC* by solving the linear system (2).
Step 3: Compute the stationary distribution π of the *DSPN* using the formula (6).
Step 4: Compute the performance measures using the formulas (18)–(24).
End

First, the *DSPN* model proposed for the queue $M/D/1/N/N$ with classical retrial policy, is validated by the exact numerical results given in (**Table 3**) [11]. We see that the performance measures corresponding the *DSPN* associated to $M/D/1/N/N$ queue are close to those obtained in [11]. Next, in **Table 4**, we

i	$p_{(0,i)}$	$p_{(1,i)}$	$\pi_{(0,i)}$	$\pi_{(1,i)}$
0	0.01010	0.03869	0.0101041	0.0386973
1	0.00074	0.09112	0.0007442	0.0911271
2	0.00078	0.15838	0.0007886	0.1583811
3	0.00081	0.21048	0.0008122	0.2104897
4	0.00070	0.21147	0.00070839	0.2114737
5	0.00048	0.15663	0.00048802	0.1566328
6	0.00025	0.08251	0.00025101	0.0825136
7	0.00009	0.02935	0.00009067	0.0293532
8	0,00002	0.00650	0.00002117	0.0065053
9	0.27×10^{-5}	0.00077	0.27×10^{-5}	0.0007788
10	0.14×10^{-6}	0.00003	$0,14 \times 10^{-6}$	0.0000362

Table 3.
Comparison of stationary distributions of the model $M/D/1/N/N$ retrial queue given in [11], with the DSPN model, "$N = 11$, $\lambda = 0.01$, $\gamma = 5.2$, and $\tau = 15$."

Measures	*DSPN* Model
n_o	8.0342602
U_s	0.6552466
$\overline{\lambda}$	1.3104932
$\overline{\gamma}$	2.0085650
$\overline{\tau}$	0.3276233
$\overline{w}$	6.1307148
ϖ	6.6307148

Table 4.
Some performance measures, "$N = 10$, $\lambda = 1$, $\gamma = 0.25$, $\tau = 0.5$."

present some performance measures. Finally, from **Figures 4–10**, we illustrate the effect of the arrival rate λ, deterministic service time τ and the retrial rate γ on the mean response time, the mean number of customers in the orbit and the server utilization.

- In **Figure 4**, we see that the mean response time of the $M/D/1/N/N$ retrial queue is maximum. The amplitude and the location of this maximum depend on the retrial rate γ. We observe that, the arrival rate λ has a significant influence on the mean response time when the retrial rate γ is weak.

- In **Figure 5**, we see that the increase in the arrival rate λ induces an increase on the mean number of customers in the orbit. We observe that, the retrial rate γ has a significant influence on the mean number of customers in the orbit. This influence is more significant for smaller value of γ.

- In **Figure 6**, we see that the increase in the arrival rate λ induces an increase on the server utilization.

- From **Figure 7**, we see that the mean response time decreases with the increase of the retrial rate γ. This decrease becomes slow with the intensifying of the secondary calls.

- In **Figure 8**, we see that the increase in the retrial rate γ induces a decrease on the mean number of customers in the orbit.

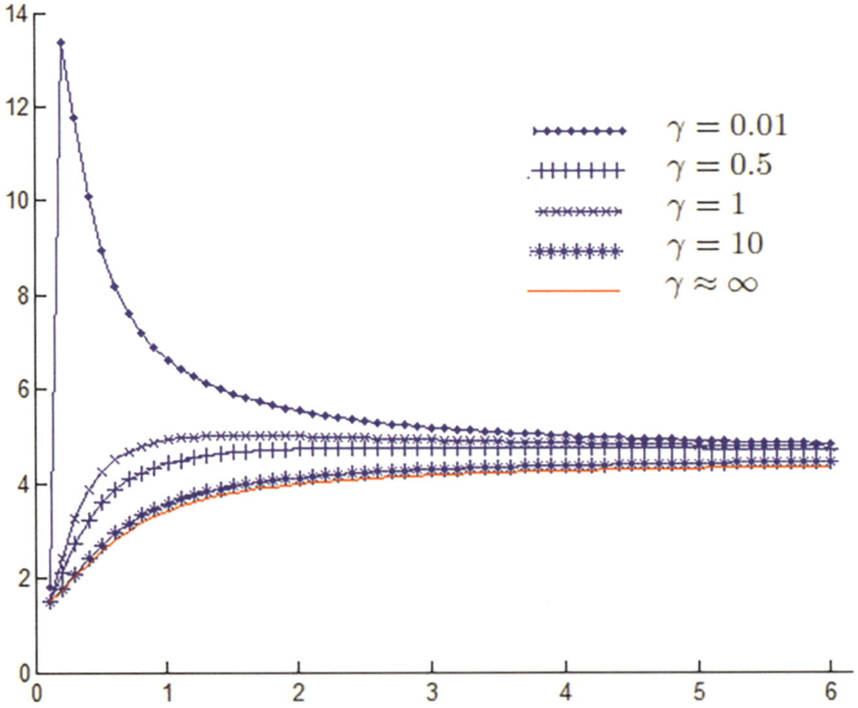

Figure 4.
Effect of arrival rate λ on mean response time ϖ, "$N = 3$, $\lambda = 10^{-3}, \ldots, 6$, $\tau = 1.5$."

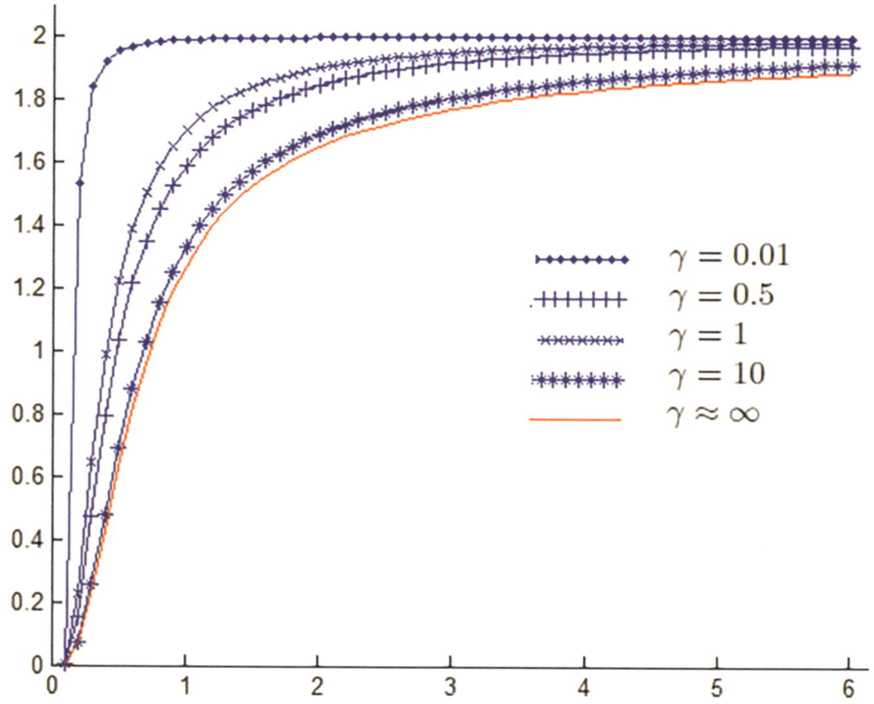

Figure 5.
Effect of arrival rate λ on mean number of customers in the orbit n_o, "$N = 3$, $\lambda = 10^{-3}, \dots, 6$, $\tau = 1.5$."

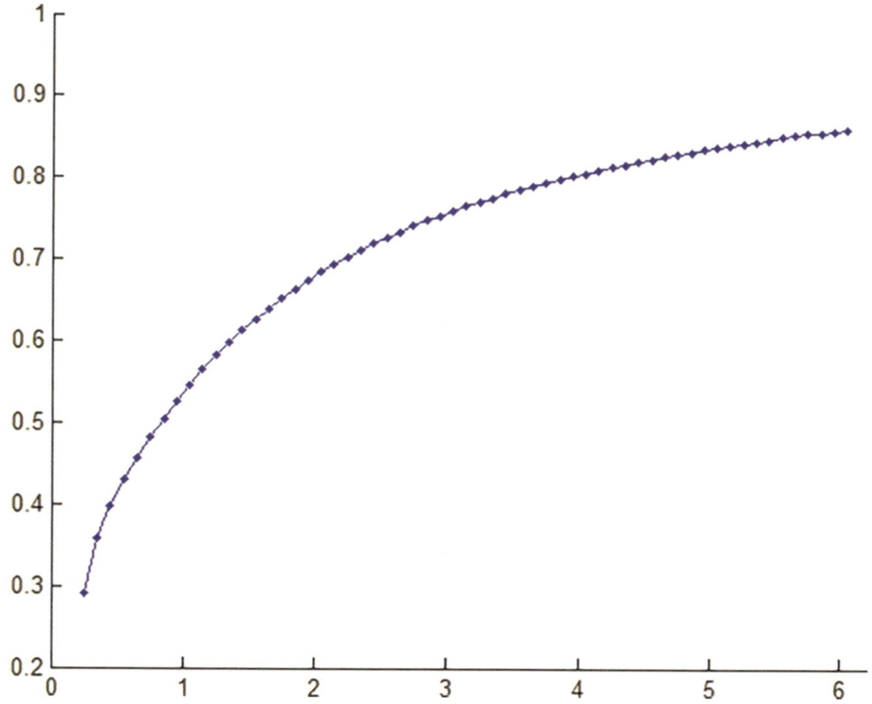

Figure 6.
Effect of arrival rate λ on the server utilization U_s, "$N = 3$, $\lambda = 0.15, \dots, 6$, $\gamma = 0.1$, $\tau = 1$."

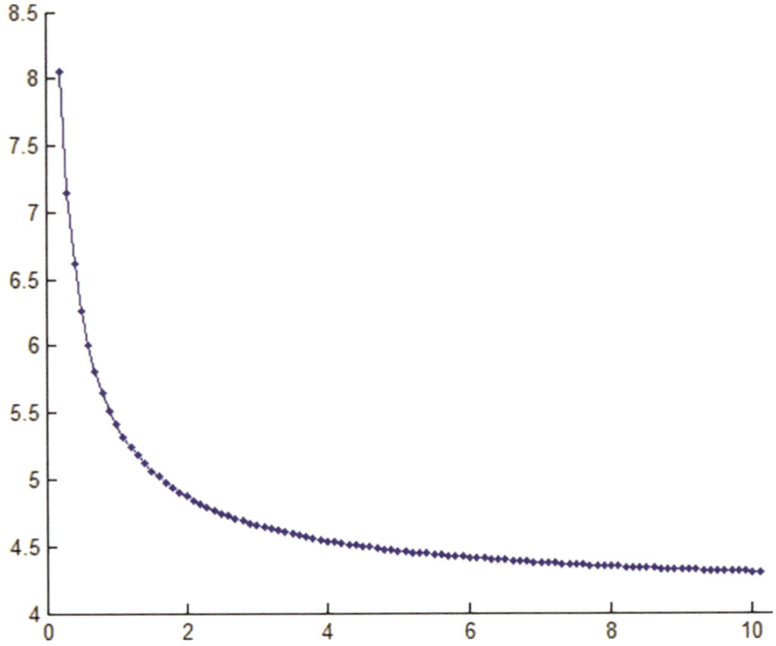

Figure 7.
Effect of retrial rate γ on mean response time ϖ, "$N = 3$, $\lambda = 0.5$, $\gamma = 10^{-1}, \ldots, 10$, $\tau = 2$."

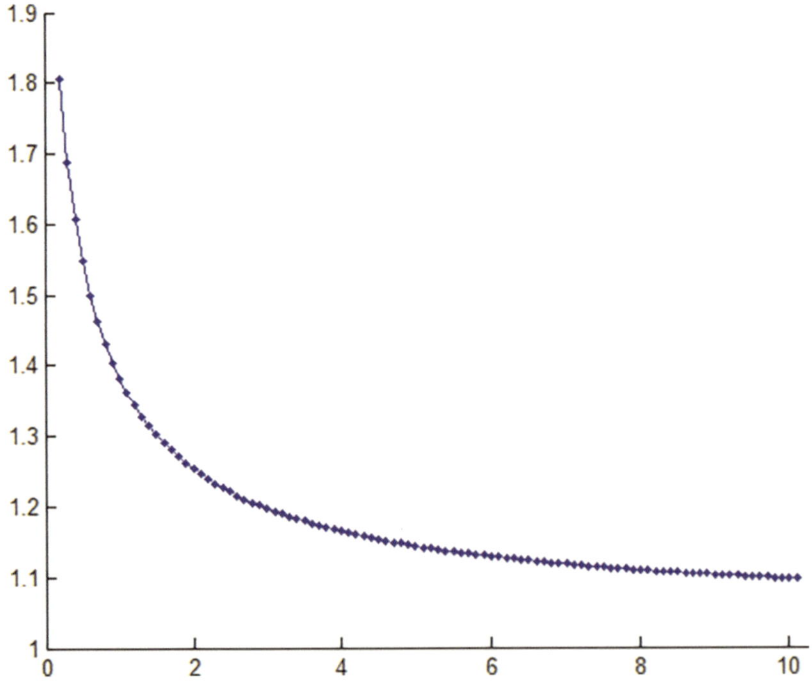

Figure 8.
Effect of retrial rate γ on mean number of customers in the orbit n_o, "$N = 3$, $\lambda = 0.5$, $\gamma = 10^{-1}, \ldots, 10$, $\tau = 2$."

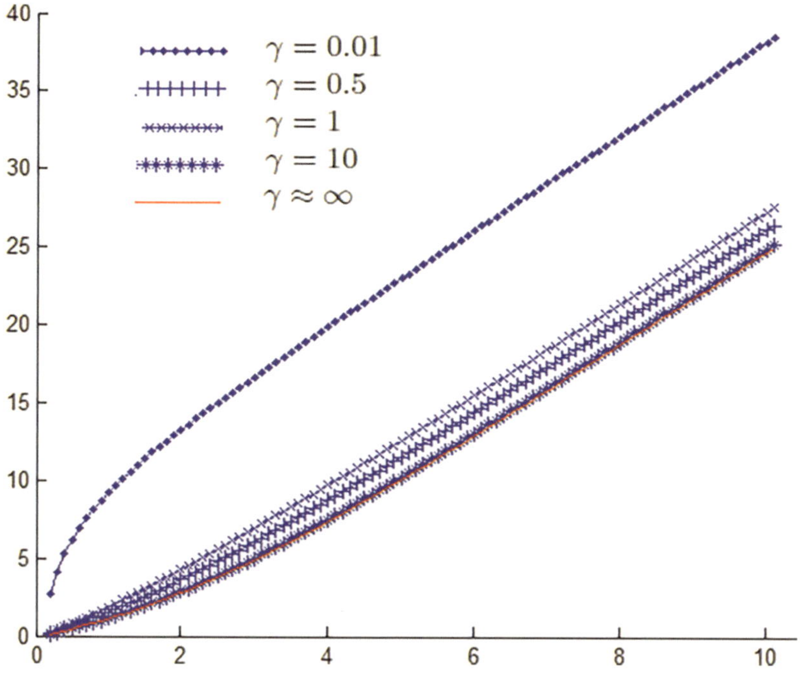

Figure 9.
Effect of service rate τ on mean response time ϖ, "$N = 3$, $\lambda = 0.2$, $\tau = 10^{-1}$, …, 10."

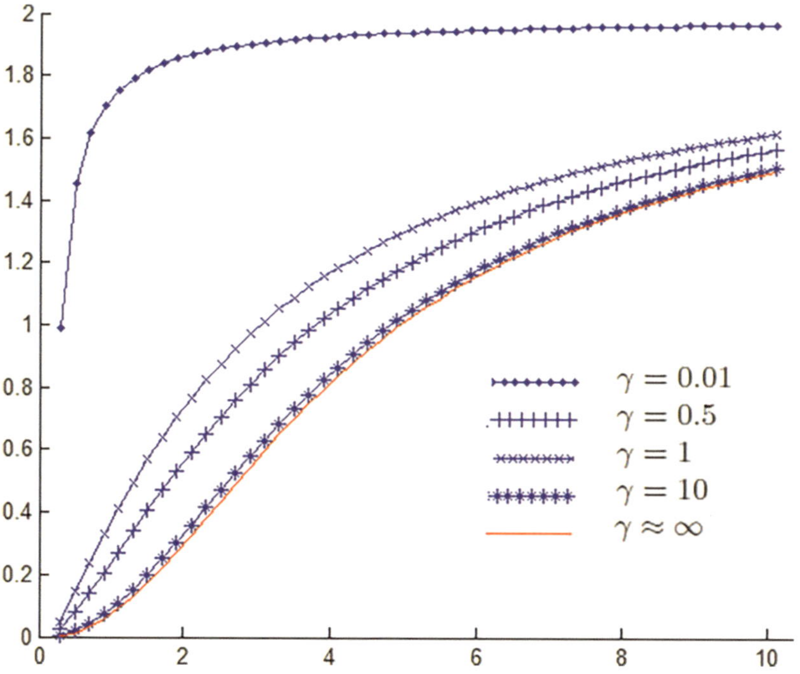

Figure 10.
Effect of deterministic service time τ on mean number of customers in the orbit n_o, "$N = 3$, $\lambda = 0.2$, $\tau = 10^{-1}$, …, 10."

- From **Figure 9**, we observe that, the deterministic service time τ has notable influence on mean response time, specifically for smaller values of τ.
- From **Figure 10**, we see that the mean number of customers in the orbit increase with the increases of the deterministic service time τ.
- From the graphs, we see that when the retrial rate tends to ∞ all the characteristics (the mean response times, the mean number of customer in the orbit, and the server utilization) approach those of the classical $M/D/1/N/N$ queue which has the lower performance bound.

8. Conclusion

We have studied a single server queue with finite sources, deterministic service, and classical retrial policy. A detailed model *DSPN* is given and validated by numerical results existing in the literature. Some performance measures of our model were given. It is very inserting to study the transient behavior and sensitivity of this system.

Author details

Lyes Ikhlef
Faculty of Sciences, Department of Mathematics, University of Algiers, Bejaia Country, Algeria

*Address all correspondence to: l.ikhlef@univ-alger.dz; ikhlefilyes@gmail.com

References

[1] Cox DR. The analysis of non-markovian stochastic processes by the inclusion of supplementary variables. Mathematical Proceedings of the Cambridge Philosophical Society. 1955; **51**(9):433-441. DOI: 10.1017/S0305004100030437

[2] Cumani A. ESP-A package for the evaluation of stochastic Petri nets with phase-type distributed transition times. In International Workshop on Timed Petri Nets. IEEE Computer Society, Washington, DC, USA. 1985. pp. 144–151

[3] Choi H, Kulkarni VG, Trivedi KS. Transient analysis of deterministic and stochastic Petri nets. The 14th International Conference on Application and Theory of Petri Nets, Chicago, U.S. A. 1993. pp. 21-25

[4] Bunday BD. Basic Queueing Theory. Australia: Edward Arnold; 1986

[5] Kashyap BRK, Chaudhry ML. An Introduction to Queueing Theory, A & A Publications. Ontario, Canada: Kingston; 1988

[6] Bhat UN. Elements of Applied Stochastic Processes. New York: John Wiley and Sons. Inc.; 1972

[7] Brun O, Garcia J. Analytical solutions of finite capacity $M/D/1$ queues. Journal of Applied Probability. 2000;**37**:1092-1098

[8] Franx GJ. A simple solution for the $M/D/c$ waiting time distribution. Operations Research Letters. 2001;**29**: 221-229

[9] Madan CK, Saleh MF. On $M/D/1$ Queue with General Server Vacations. Information and Management Sciences. 2001;**12**(2):25-37

[10] Choi H, Mainkar V, Trivedi KS. Sensitivity analysis of deterministic and stochastic Petri nets, Inetrnational Workshop on Modeling, Analysis and Simulation of Computer and Telecommunication Systems, San Diego, USA. 1993. pp. 17-20

[11] Artalejo JR, Gomez-Corral A. Information theoretic analysis for queueing systems with quasi-random input. Mathematical and Computer Modelling. 1995;**22**:65-76

[12] Wu X, Ke X. Analysis of an $M/\{Dn\}/1$ retrial queue. Journal of Computational and Applied Mathematics. 2007;**200**:528-536

[13] Artalejo JR. Accessible bibliography on retrial queues: Progress in 2000-2009. Mathematical and Computer Modelling. 2010;**51**(9–10):1071-1081

[14] Falin GI. A survey of retrial queues. Queueing Systems. 1990;**7**:127-168

[15] Falin GI, Templeton JGC. Retrial Queues. London: Chapman and Hall; 1997

[16] Yang T, Templeton JGC. A survey on retrial queue. Queueing Systems. 1987;**2**: 201-233

[17] Marsan MA, Chiola G. On Petri nets with deterministic and exponentially distributed firing times. In: Rozenberg G, editor. Advances in Petri Nets 1986, Lecture Notes in Computer Science. Vol. 266. Springer-Verlag; 1987. pp. 132-145

[18] Ciardo G, German R, Lindemann C. A characterization of the stochastic process underlying a stochastic Petri net. IEEE Transactions on Software Engeneering. 1994;**20**:506-515

Chapter 3

Markov Models Related to Inventory Systems, Control Charts, and Forecasting Methods

Sivasamy Ramasamy and Wilford B. Molefe

Abstract

A Markov model describes a randomly varying system that satisfies the Markov property. This means that future and past states at any time are independent of the current state. The two most commonly used types of Markov models are Markov chains and higher-order Markov chains. Therefore, three types of Markov models are proposed in this chapter of the book: (i) supply chain management, (ii) Markov queue waiting time monitoring, and (iii) Markov fuzzy time series forecasting. The introduction introduces Markov chain (MC) and summarizes the most important aspects of Markov chain analysis. The first model explores a Markov queue model coupled to a storage system using the classical (0, Q) policy. The second model focuses on the M/M/1/N service mode and develops a control chart for an $M_s/M_s/1/N$ type simulated queue to monitor customer waiting times. The third is a higher-order Markov model (HOMM), which uses fuzzy sets to predict future states based on given hypothetical time series data. Numerical calculations are designed to find optimal order quantities, monitor customer wait times, and predict future HOMM conditions.

Keywords: Markov chain, queue inventory, control chart, probability vector, rainfall

1. Introduction

The term probability refers to the proportion of interesting events observed in one or more random experiments. The probability of an event A occurring in an experiment is given by the ratio f/T, where f = A can occur and T = the total number of possible outcomes of this experiment. We denote the set of all outcomes of a random experiment by $\Omega = \{\omega\}$, which is called the sample space. A real-valued function $X = X(\omega)$ that gives a unique real value for each term "ω" in the real set R (= $(-\infty, \infty)$) is called a (real valued) random variable. We classify a random variable as discrete if it has only a few specific values, such as −1, 3, 8, and 17. Otherwise, it is classified as continuous. A family of random variables indexed by time "t" defined on a state space $\mathbb{S}$ i.e., $X = \{X(t) \in \mathbb{S} : t \in T = [0, \infty)\}$ is called a stochastic process. Thus, a typical Stochastic process is indexed by time "t" and takes its state values over a space $\mathbb{S}$ is represented as

$$X = \{X(\omega, t) \in \mathbb{S}, \omega \in \Omega, t \in T\} \tag{1}$$

IntechOpen

1.1 Continuous time (CT) Markov process (MP) $\{X(t), t > 0\}$

If the probability of future state of $X(t)$ depends only on the present but not on the past states then $\{X(t), t > 0\}$ is a MP for any set of time points $t_1 < t_2 < \cdots < t_{n+1}$ and any set of states $x_1 < x_2 < \cdots < x_{n+1} \Rightarrow$

$$\Pr(X(t_{n+1}) = x_{n+1} | X(t_1) = x_1, X(t_2) = x_2, \ldots, X(t_n) = x_n) = \Pr(X(t_{n+1}) = x_{n+1} | X(t_n) = x_n). \tag{2}$$

1.2 Markov Chain $\{X_n: n = 0, 1\ 2, \ldots\}$

- A discrete time MP $\{X_t = X_n: n = 0, 1\ 2, \ldots\}$ is called a Markov chain (MC). Let the conditional probability of moving from state i to state j be $P_{ij} = P(X_{(n+1)} = j \mid X_n = i)$, $n \geq 0$ and i, and j = 0,1,2, … .

- Define one step transition probability matrix (TPM): $\mathbb{P} = [P_{ij}]$ Let $P_{ij}^{m} =$. The conditional probability of moving to state j starting from state i in m steps which can be calculated using the Chapman Kolmogorov (CK) equations:

$$P_{ij}^{m} = \Pr(X_m = j | X_0 = i) = \sum_{k=0}^{\infty} P_{ik}^{r} P_{kj}^{m-r} \tag{3}$$

$$\text{for } i, j = 0, 1, 2, \ldots \text{ and } m \geq r \geq 0. \tag{4}$$

- Thus, the mth step TPM is $\mathbb{P}^m = \left[P_{ij}^{m}\right]$.

- A MC is completely determined if the initial distribution $P(X_0 = j)$ and its one step TPM P are all known or estimated by conducting a sample survey.

1.3 Classification of States of a MC

- State j is *accessible* from state i if there exists an integer m such that $P_{ij}^{m} > 0$.

- Two states i and j *communicate* with each other if both are accessible from each other.

- The TPM of an MC defined on a state space $\mathbb{S}$ is said to be *irreducible* if all states communicate with each other state in $\mathbb{S}$.

- State i is *recurrent* if the system will return to it after some finite time in the future.

- If a state is not recurrent, it is *transient*.

- A state is *periodic* if it can only return to itself after a fixed number of transitions greater than 1 (or multiple of a fixed number).

- A state that is not periodic is *aperiodic*.

- An *absorbing* state is one that locks in the system once it enters.
- All states belonging to a class are either recurrent or all transient and may be either all periodic or aperiodic.

1.4 Stationary distribution of an aperiodic and irreducible MC

- Let and

$$\pi_j = \lim_{n\to\infty} P_{ij}^n \text{ and } \Pi = \{\pi_0, \pi_1, \pi_2, \dots\}, \sum_{j=0}^{\infty} \pi_j = 1. \tag{5}$$

- Under certain conditions, these limiting probabilities π_j can be shown to exist and are independent of the starting state i.
- They represent the long run proportions of time that the process spends in each state j. Also π_j = the steady-state probability that the process will be found in each state j. Computation of the row vector Π:

$$\Pi\,\mathbb{P} = \Pi, \quad \sum_{j=0}^{\infty} \pi_j = 1. \tag{6}$$

- For an aperiodic and recurrent MC on the state space $\Omega = \{1, 2, \dots, m\}$. Let $\pi_j = \lim_{n\to\infty} P_{ij}^n$, $\sum_{j=0}^{m} \pi_j = 1$ and row vector $\Pi = (\pi_0, \pi_1, \dots, \pi_m)$.
- Then, the stationary probability vector Π satisfies the following set of simultaneous equations:

$$\Pi\,\mathbb{P} = \Pi, \quad \Pi\,\mathbf{1} = 1 = \sum_{j=0}^{m} \pi_j \tag{7}$$

- **Computational Hint:** One of the m equations of $\Pi\,\mathbb{P} = \Pi$ is redundant. It is advised that replacing any of the equations of $\Pi\,\mathbf{P} = \Pi$ by $\Pi\,\mathbf{1} = 1$, we form a set of m simultaneous equations which yield the unique Π. One can find quite a number of applications of finite state MCs in our social life activities.

1.5 Generator Matrix of a continuous time MP

Let T_{ij} denote the time for the MP to go from state i to state j and q_{ij} be the known rate of transition from state i to state j.

Then $T_{ij} \to e^{q_{ij}}$. Let $q_i = \sum_{j \neq i} q_{ij}$,

The infinitesimal generator is

$$Q = \begin{bmatrix} -q_0 & q_{01} & q_{02} & q_{03} & \cdots \\ q_{10} & -q_2 & q_{12} & q_{13} & \cdots \\ q_{20} & q_{21} & -q_3 & q_{23} & \cdots \\ \vdots & \vdots & \vdots & \vdots & \vdots \end{bmatrix} \tag{8}$$

The stationary distribution is obtained by solving the system of equations

$$\Pi Q = \mathbf{0}, \quad \Pi\mathbf{1} = 1 = \sum_{n=0}^{\infty} \pi \tag{9}$$

Basic information on MC definition, classification of states, etc. can be found in [1, 2]. Instructions for calculating the state characteristics of different types of MCs using software "markov chains" are briefly discussed in [3].

1.6 Ordering Markov models discussed

Section 2 considers customer arrivals to demand a single product of supply chain management. All arrivals join an M/M/1/N/FCFS system attached to a storage house under the classical (0, Q) policy. The joint process of limiting queue length and the inventory level is denoted by $\mathbf{Z} = \{(\mathrm{L}, \mathrm{J})\} = lim_{t \to \infty}\{(L(t), J(t))\}$. We then obtain stationary probability vector of the **Z** process. A cost function is formed as an implicit function of the order quantity Q and other input values, and the optimal order quantity that minimizes its total cost is determined.

Section 3 presents the analysis on an M/M/1/N queue and develops simulated queue $\mathrm{M}_s/\mathrm{M}_s/1/\mathrm{N}$. The steady state characteristics such as the embedded queue length and waiting times at all successive transition epochs of the $\mathrm{M}_s/\mathrm{M}_s/1/\mathrm{N}$ system are obtained and a control chart is constructed to monitor the impacts of the customer's waiting time distribution.

Section 4 introduces a HOMM for analyzing a fuzzy time series as a data sequence. The main focus is to build a first-order MC (FOMC) and another second-order MC (SOMC) for the same data sequence under study. The prediction error (PE) is calculated for both FOMC and SOMC and a comparison is made to isolate the best performer. Section 5 provides a formal concluding report.

2. Markov model for supply chain managements

A Markov chain sequence can be thought of as a stochastic model. Such a model involves mathematical considerations to estimate the probability of a sequence of future states based on the most recent event rather than past history. The Markov model is a stochastic method for randomly varying systems with the Markov property. This means that at any moment the next state depends only on the current state and is independent of the past.

Markov analysis is a method for predicting the value of a variable whose predicted value is affected only by its current state and not by any past activity. Basically, it predicts a random variable based only on current conditions. The three simple forms of Markov models are (i) MC, (ii) Markov decision process (MDP), and (iii) HOMM. Each model has applications in queuing theory, inventory systems, and reliability theory.

To learn the analysis of queuing problems, researchers can read [1, 2, 4–9]. Regarding stochastic model building using queuing storage systems, you can refer to [8, 10–12]. The use of matrix analysis to find steady state results for queue-length processes is well covered in the book by [13].

2.1 Markov decision processes of a supply chain

Supply chain (SC) is a system that links sales branches into a supply chain that would financially improve customer satisfaction and loyalty. Several SC researchers have used integrated queues with inventory policies to reduce lead time and inventory costs.

This section discusses a Markov model that tracks inventory levels in a warehouse connected to a queuing system that sells individual products based on customer demand. This is called the Queue inventory model (QIM), which allows demands to be made during the inventory period. However, those customers who arrive when the warehouse is empty will not be processed until after the ordered products arrive.

The proposed SC facility delivers either a package of $Q > 0$ fixed-size items for the normal order or only one item for an emergency order to a warehouse that processes customer requirements through a Markov service node M/M/1/N.

Denoting the number of customer demands occurred at time "t" by X(t) and the present on-hand inventory level by Y(t), we propose matrix methods to study the bivariate stochastic process **Z**(t) = (X(t), Y(t)) defined on the state space **E** = $N_0 x Q_0$ where N_0 = 0, 1, 2, ... , N and Q_0 = 0, 1, 2, ... , Q. The system has a single server who admits a maximum number "N" of customers and does not admit those customers who arrive at instances when X(t) = N called lost customers. The queue discipline is on a first-come, first-served basis (FCFS).

Customers requesting one unit of stock arrive at a node of a Poisson process with mean parameter (λ) when X(t) = n, n*ge*0. For each request, after an exponential service time, one product unit is offered with a state-dependent parameter μ_n. Here we have two types of inventory ordering policies called "regular" and "emergency":

1. The normal ordering is the classical (0, Q(>1)) policy, which is set whenever the MDP **Z**(t) moves from (n,0) to (n + 1,1) $n \geq 1$. The lead time is assumed to be an exponentially and identically distributed random variable with rate parameter γ.

2. The emergency ordering type is the classical (0, Q(=1)), which triggers an order for one product whenever the **Z**(t) process moves from (1,1) to (0,0). It is assumed that the emergency fulfillment time is a small interval of unit length (IUL) and the urgent order is replenished at a rate of "1." This emergency order period expires at the end of the IUL. In this particular period, P [(0,0) $\rightarrow$ (0,1)] = $1/(\lambda + 1)$, P[(0,0) $\rightarrow$ (1,0)] = $\lambda/(\lambda + 1)$, that is, whether the replenishment of one product or the arrival of a customer demand is unavoidable.

3. Our M/M/1/N service system will not accept every incoming customer if there are N customers in front of the system. There can be at most one pending order where "0" is the order point and the order quantity is at most Q units per replenishment for the normal type and only one product for the emergency type.

Due to the properties of exponential distributions and the mixed-type of normal and emergency practices, the proposed sequence **Z**(t) forms an aperiodic and positive recurrent Markov decision process (MDP). The set properties of the **Z** system can be easily checked at each stage of service completion. Our main objective is to use a linear function of expected total cost to determine the optimal order quantity and order patterns.

2.2 Generator matrix of the MDP process Z

It is interesting to note that the MDP process **Z** is also a level-dependent quasi-birth-and-death (LDQBD) in continuous time. We have the (N + 1) level L_n such that $\bigcup_{n=0}^{N} L_n = E$. The infinitesimal generator matrix **G** has the tri-diagonal structure and each submatrix of **G** is a square matrix of order (Q + 1):

$$\mathbf{G}=\begin{bmatrix} B & \mathbf{A_0} & 0 & 0 & \ldots & \mathbf{0} & \mathbf{0} \\ A_2 & A_1 & \mathbf{A_0} & 0 & \ldots & \mathbf{0} & \mathbf{0} \\ 0 & A_2 & A_1 & \mathbf{A_0} & \ldots & \mathbf{0} & \mathbf{0} \\ 0 & 0 & A_2 & A_1 & \ddots & \mathbf{0} & \mathbf{0} \\ \vdots & \vdots & \vdots & \ddots & \ddots & \mathbf{A_0} & \mathbf{0} \\ \mathbf{0} & \mathbf{0} & \mathbf{0} & \mathbf{0} & \ddots & A_1 & \mathbf{A_0} \\ \mathbf{0} & \mathbf{0} & \mathbf{0} & \mathbf{0} & \ldots & A_2 & (A_1+\mathbf{A_0}) \end{bmatrix} \tag{10}$$

Each sub-matrix of the **G** is a square matrix of order (Q + 1) indexed by j = 0, 1, …, Q. The precise structures of $\mathbf{B} = (b_{i\,j})$, $\mathbf{A}_0 = \left(a_{i\,j}^{(0)}\right)$, $\mathbf{A}_1 = \left(a_{i\,j}^{1)}\right)$, and $\mathbf{A}_2 = \left(a_{i\,j}^{(2)}\right)$ are now described:

$$b_{i\ j}=\begin{cases} b_{1\quad 1}=-(\lambda+1)\quad b_{1\quad 2}=1\ b_{1\ j}=0\quad \text{for}\quad j=3,\ 4,\ldots,\ (Q+1) \\ b_{i\ j}=-\lambda\ \text{for}\ 1=j=2,\ ..\ (Q+1) \\ 0\quad otherwise \end{cases} \tag{11}$$

$$a_{i\ j}^{(0)}=\begin{cases} \lambda\ \text{for}\ i=j=1,\ 2,\ ..\ (Q+1) \\ 0\ otherwise \end{cases} \tag{12}$$

$$a_{i\ j}^{(1)}=\begin{cases} a_{1\quad 1}^{(1)}=-(\lambda+\nu)\quad a_{1\quad (Q+1)}^{(1)}=\nu \\ -(\lambda+\mu)\ \text{for}\ i=j=2,\ 3,\ \ldots,(Q+1) \\ 0\quad otherwise \end{cases} \tag{13}$$

$$a_{i\ j}^{(2)}=\begin{cases} a_{i\quad (i-2)}^{(2)}=\mu\quad \text{for}\quad i=2,\ \ldots,\ (Q+1) \\ 0\quad otherwise \end{cases} \tag{14}$$

The solution of matrix equations generated by level-dependent QBD processes has been widely studied by several researchers, see [6]. A useful matrix analysis method for computing the steady-state probability distributions of the queue-length process of QBDs is presented in Ref. [13].

Distribution: Denote by $\mathbf{\Pi} = (\boldsymbol{\pi}_0, \boldsymbol{\pi}_1, \boldsymbol{\pi}_2, \ldots, \boldsymbol{\pi}_N)$, the steady state distribution of **Z** and by $\boldsymbol{\pi}_n$ the row vector $\boldsymbol{\pi}_n = (\pi(n, 0), \ldots, \pi(n, Q))$ associated with level L_n, $n \in N_0$. Let **e** = (1,1, … ,1) be a column vector of unit elements.

We now discuss a simple linear level reduction method in two phases for a positive recurrent case that leads to computation of the limiting distribution **Π**: **Phase 1**: Iteratively reduce the state space from the level "N" by removing one level at each step until we reach the level "0" and check if the generator of the level "0" corresponds to a Markov process.

Compute, the $U^{(n)}$ matrices and the rate matrices R_n in terms of the sub-matrices of the generator **G**:

$$U^{(N)} = A_1^{(N)} + A_0, \tag{15}$$

$$U^{(n)} = A_1^{(n)} - A_0\left[U^{(n+1)}\right]^{-1} A_2^{(n+1)} \quad \text{for} \quad n=(N-1),\quad (N-2),\ \ldots,\ 1 \tag{16}$$

$$U^{(0)} = B - \mathbf{A_0}\left[U^{(\mathbf{1})}\right]^{-1} A_2{}^{(\mathbf{1})} \tag{17}$$

$$\mathbf{R}_n = -\mathbf{A}_0 \left[\mathrm{U}^{(n)}\right]^{-1}, \quad n = 1, 2, \ \ldots, N \tag{18}$$

For $1 \leq n \leq N$, the matrix $U^{(n)}$ records the transition rate from level L_n to L_n before a visit to $L_{(n-1)}$. In addition, the matrix $-\left[U^{(n)}\right]^{-1} A_2^{(n)}$ records the first passage time probabilities from level L_n to L_n (**Figure 1**).

Phase 2:

Theorem 1.1 The unique stationary joint distribution vector **Π** of the Queue length and the inventory level of the **Z = (X, Y)** process is given by

$$\pi_n = \pi_{n-1}\ \mathbf{R}_n \ for \quad n = 1, 2, \ldots, N \quad and \quad \mathbf{\Pi e} = 1. \tag{19}$$

2.3 Alternative way of computing the stationary probability vector Π

Phase I: Compute.

$$\mathbf{C}_0 = \mathbf{B}, \mathbf{C}_i = \mathbf{A}_1^{(i)} + \mathbf{A}_2^{(i)}(-\mathbf{C}_{i-1})^{-1}\mathbf{A}_0 \text{ for } i = 1, 2, \ldots, (N-1), \text{and} \tag{20}$$

$$\mathbf{C}_N = \left(\mathbf{A}_1^{(N)} + \mathbf{A}_0\right) + \mathbf{A}_2^{(N)}\left(-\mathbf{C}_{(N-1)}\right)^{-1}\mathbf{A}_0 \tag{21}$$

It is noticed that the matrix $\left(-\mathbf{C}_i^{-1}\mathbf{A}_0\right)$ records the first passage probabilities from level L(i) to L(i + 1).

Phase II: The Stationary probability vector $\mathbf{\Pi} = (\mathbf{\Pi}_0, \ \mathbf{\Pi}_1, \ \mathbf{\Pi}_2, \ \ldots, \mathbf{\Pi}_N)$, $\Sigma_{n=0}^{N}\mathbf{\Pi}_n \ e = 1$ is determined by

$$\mathbf{\Pi}_N C_N = 0, \tag{22}$$

$$\mathbf{\Pi}_n = \mathbf{\Pi}_{n+1} A_2^{(n+1)}(-C_n)^{-1} \quad \text{for} \quad n = \text{(N-1), (N-2), } \ldots, 3, 2, \ 1 \ \text{ and } 0 \tag{23}$$

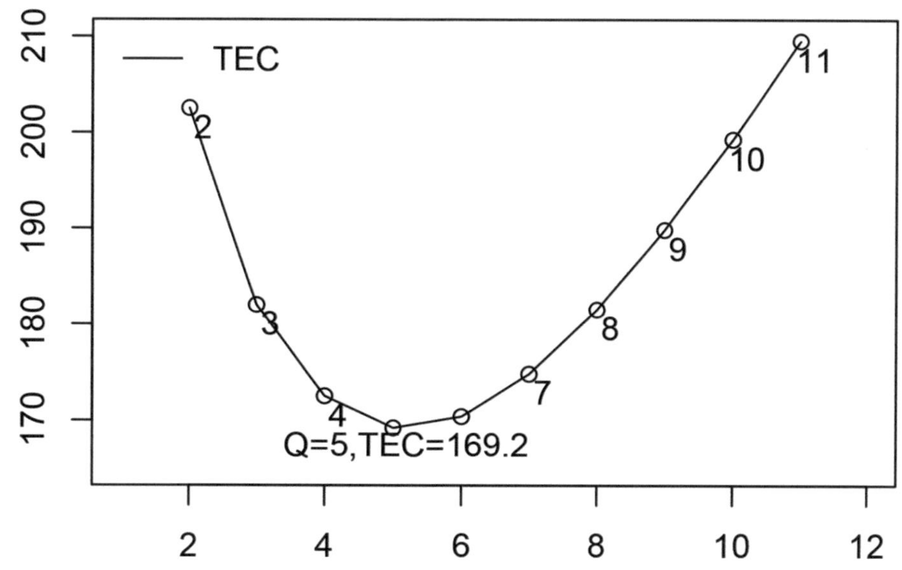

Figure 1.
Q versus TEC.

2.4 Useful metrics derived from Π

The marginal distribution, say $\mathbf{\Pi}_x$, of the queue length process X and that of the inventory process, say $\mathbf{\Pi}_y$ are now obtained from the joint probability vector $\mathbf{\Pi}$ respectively by $\mathbf{\Pi}_x = (x_0, x_1, \dots, x_N)$ where $x_n = \pi_n \mathbf{e}$, for n = 0,1, … , N and $\mathbf{\Pi}_y = (y_0, y_1, \dots, y_Q)$ where $\left(y_j = \Sigma_{n=0}^{N} \ \pi(n,j)\right.$ for j = 0, 1, … ,Q.

Mean system Size L and Mean wait $\overline{W}$: The mean system size L and the mean sojourn time $\overline{W}$ of the queue length process X are given by

$$L = \sum_{n=0}^{N} n\pi_n \quad and \quad \overline{W} = \frac{L}{\lambda_{eff}}, \ \lambda_{eff} = \lambda(1 - \pi_N) \tag{24}$$

Mean Inventory Size $\overline{V}$ and Mean Reorder Rate $\overline{R}$: The mean inventory size $\overline{V}$ and the mean reorder rate $\overline{R}$ of the inventory level process Y are given by

$$\overline{V} = \sum_{j=0}^{Q} j \ y_j \quad and \quad \overline{R} = \mu \ \pi(1,1) + \sum_{n=2}^{N} \mu \quad \pi(n, \ 1) \tag{25}$$

Total expected cost (TEC) Let us consider three different costs i.e., h = holding cost per item per unit of time, c1 = ordering cost per order and c2 = waiting time cost per unit of time. A linear TEC function is then designed to determine an optimum order quantity Q* for which TEC(Q*) is the minimum:

$$TEC = h \ \overline{V} + c1\overline{R} + c2\overline{W} \tag{26}$$

2.5 Determination of optimum order quantity, Q*, N = 25

A numerical representation is now considered to determine the optimal order quantity that minimizes the TEC described in (14). We set the input dataset as follows: N = 25, arrival rate λ = 2.2, service amount μ = 2.5 and replenishment rate ν = 2.25, and holding cost h = $ 27.5 order price c1 = $ 8.4, and waiting time costs c2 = $ 12.

In this case, the TEC values for varying order quantity Q are calculated with values 2 to 11, and the same is shown graphically in **Figure 1**. The order quantity that minimizes TEC is Q* = 5 and the corresponding price is $ 169.2.

3. Markov modeling of queues with finite capacity: M/M/1/N

The chapter considers a simple simulated Markov model of a single server M/M/1/N. It focuses on a process control method that can monitor customer waiting times. Any service counter with a buffer of waiting customers naturally leads to a limited queue.

3.1 Simulation of cumulative arrival and departure epochs

The formal simulation method is now carried out to fit the M/M/1/N model, and the result is called the simulated Markov model $M_s/M_s/1/N$. Let the arrival rate of the

M/M/1/N exponential population under investigation be λ and the service level be μ. We take enough samples from each exponential population to simulate consecutive customer arrival and departure periods, the cumulative arrival times, the number of customers in such periods, and the waiting time of each customer for a given number of arrivals such as M = 1000. The simulation process algorithm:

Step 1: Simulate "M" number of inter-arrival times $\{a_1, a_2, \ldots, a_M\}$ from the exponential distribution with parameter λ. Let the starting time of the simulation process be zero, i.e., $T_0 = 0$. Then calculate the sequence of cumulative arrival times $T_j = T_{(j-1)} + a_j$ for j = 1, 2, …, M.

Step 2: Simulate "M" number of service times $\{s_1, s_2, \ldots, s_M\}$ from the exponential distribution with parameter μ. Then calculate the sequence of service start time times $\{S_j\}$ and the sequence of departure epochs $\{D_j\}$: $S_1 = (T_1)$, $D_1 = S_1 + s_1$.

$S_j = \text{Max}\{(T_j), D_j)\}$ and $D_j = S_j + s_j$ for $j = 2, 3, \ldots, M$.

The waiting time W_j in the system for the jth arrival is then given by $W_j = D_j - T_j$.

Step 3: The (pre-arrival) number $L_j^{(a-)}$ of customers present at the arrival points $\{T_j\}$ and post-departure number $L_j^{(d)}$ of customers present at the departure points $\{D_j\}$ are then calculated using appropriate formulae during the simulation period.

Step 4: If the maximum value of the sequence $\left\{L_j^{(a-)} : j = 1, 2, \ldots, M\right\}$ is N-1, then the maximum capacity of the simulated Markov model is $M_s/M_s/1/N$. For cross checking, verify the fact that the maximum value of the sequence $\left\{L_j^{(d)} : j = 1, 2, \ldots, M\right\}$ also equals to N-1.

Step 5: Let f_k = Number of times the event $L_j^{(a-)} = k$ occurred for k = 0, 1, …, (N-1) over the arrival epochs $\{T_j\}$ for j = 1, 2, …, M. Then $\Sigma_{k=0}^{(N-1)} f_k = M$ Thus, the prearrival epoch queue length distribution is given by $\left\{\pi_k^{(a-)} = \frac{f_k}{M}\right\}$..

Step 6: Let F_k = Number of times the event $L_j^{(d)} = k$ occurred for k = 0, 1, …, (N-1) over the departure epochs $\{D_j\}$ for j = 1, 2, …, M, where $\Sigma_{k=0}^{(N-1)} F_k = M$. Thus, the departure epoch queue length distribution is given by $\left\{\pi_k^{(d)} = \frac{F_k}{M}\right\}$. For a cross-checking, verify the fact that $f_k = F_k$ for k = 0, 1, 2, …, (N-1).

3.2 How to fit M/M/1/N

The special feature is that the maximum number "N" of the simulated version $M_s/M_s/1/N$ plays a crucial role. Let us denote the estimated results obtained from the above simulated outcomes for pre-arrival epoch queue length probabilities by $\pi_j^{(a-)}$ and post-departure epoch probabilities by $\pi_j^{(d)}$. Then it can be shown that

$$\pi_j^{(a-)} = \pi_j^{(d)}; \quad j \in \mathbb{S}^{(a-)} = \{0, \ 1, \ 2, \quad \ldots, \ (N-1)\} = \mathbb{S}^{(d)} \tag{27}$$

Let $\left\{t_0^{(a)} = 0, t_n^{(a)}\right\}$ be the sequence of times at which $Y\left(t_n^{(a)}\right)$ represents the queue length immediately after the nth arrival epoch $t_n^{(a)}$ on the state space $\mathbb{S}^{(a)} = \{1, 2, \ldots, N\}$. Let, $\pi_j^{(a)} = P\left(Y\left(\binom{(a)}{n}\right) = j\right)$:

$$\pi^{(a)}_{(j+1)} = \pi^{(a-)}_j; \quad j \in \mathbb{S}^{(a-)} \tag{28}$$

Embedded queue length (EQL) process $\mathbf{Y} = \{Y_n \in \mathbb{S} = \{0, 1, 2, \ \ldots, N\}\}$: Let us now describe the embedded queue length Y_n found at the nth transition epoch t_n. Thus, t_n represents the nth order statistics of the union of $\{t_n^{(a)}\} \cup \{t_n^{(d)}\}$. Let π_j = P() Y_n = j) for $j \in \mathbb{S} = \{0, 1, 2, \ \ldots, N\}$.

$$\pi_j = \begin{cases} \pi_0^{(d)}/2 \\ \left(\pi_j^{(d)} + \pi_{j+1}^{(a)}\right)/2 \ \text{for} \quad j = 1, 2, \ ..(N-1) \\ \pi_N^{(a)}/2 \quad \text{for} \quad j = N \end{cases} \tag{29}$$

Some specific applications of control chat methods are discussed in [5, 14] (**Figure 2**).

Detailed information that explores applications of HOMM and data sequence analysis using fuzzy sets are found in [15–18]. Theory of Fuzzy sets was introduced by Zadeh [19].

3.3 Numerical evaluation of a simulated Ms/Ms/1/N queue

Now, with a simulation study of M = 1000 patients, we try to perform the queuing functions of a health care unit in a way that ensures the best results of an $M_s/M_s/1/N$ queue. The processing time per customer is exponentially distributed with an average rate of μ = 0.27. Patients arrive randomly, with an average of one position every 4 minutes, and are treated on a first come, first served basis. The simulation route is run with an exponential arrival rate λ = 0.25 and an exponential service μ using r-code to implement the algorithm described above.

The rth raw moment of the EQL is given by $E(Y^r) = \Sigma_{j=0}^{N} j \ \pi_j$.

The estimated values of the all 2000 (= 1000 arrival point and 1000 departure epochs) transition epochs t_n: n = 1,2, ..., 2000 and the corresponding waiting time $W_n = L_n/\lambda$ of patients are calculated.

A box-and-whisker diagram of the waiting time distribution W_n: n = 1,2, ..., 2000 and its histogram are displayed side by side in **Figure 2**. As expected, the box-and-

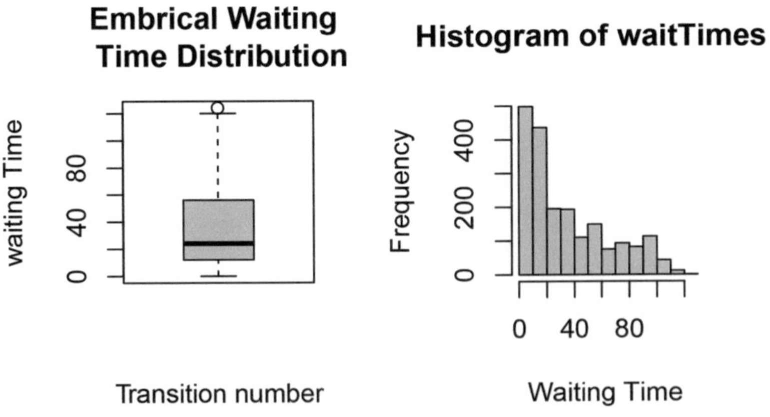

Figure 2.
Box plot and histogram of waiting times.

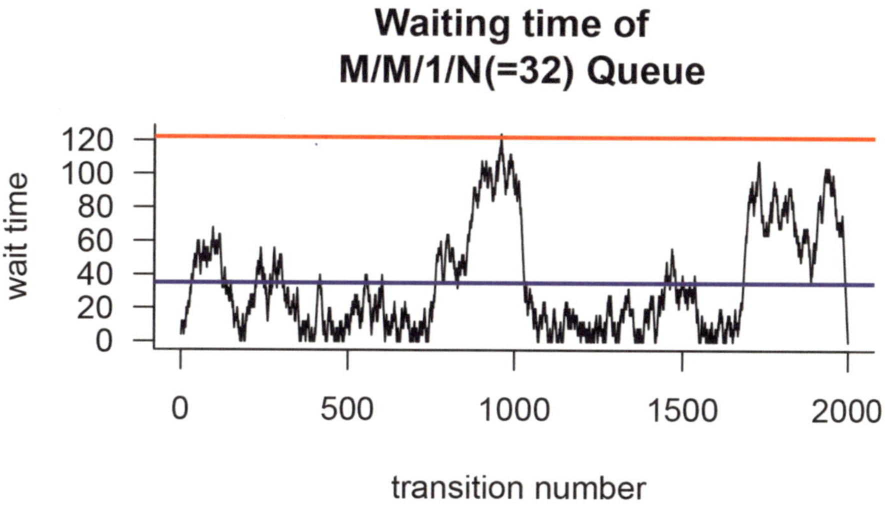

Figure 3.
Transition epochs vs. waiting times.

whisker plot shows the sizes, spread, and skew of the interquartiles (Q1 = 12, Q2 = 24, Q3 = 56), max = 124, and whiskers (bottom = 0, top = 122) outside the upper and lower quartiles of waiting times. In addition, one exterior remains above the upper whisker, i.e., max W_n: n = 1,2, ..., 2000 = 124. Using these facts together with some facts revealed by the histogram, we can conclude that the waiting time distribution can follow, approximately, an exponential distribution with a long tail to the right. A control chart (**Figure 3**) is designed to show additional information of the pair sequence $\{(t_n, W_n): n = 1,2, \ldots, 2000\}$.

Any "health" management can refer to the **Figure 3** and make certain decisions. First, management must maintain all types of resources necessary to treat and accommodate up to N = 32 patients, such as beds, medical equipment. Second, management can inform patients that the average waiting time is about 35 minutes and the maximum waiting time is about 124 minutes. If the administration installs an additional server during the transition periods of 995 and 1005, the patient's waiting time can also be reduced. So this section explores how simulation technology can be used to build Markov-type queuing models and use it to better serve ordinary people.

4. Higher-order Markov model (HOMM)

This section discusses different characteristics of an higher-order Markov model (HOMM). The purpose is to explore briefly how a HOMM can be used in forecasting of time series observed over agricultural productions, stock market prices, oil demand, etc. The first step is to covert the given time series into a categorical sequence by fuzzifying the given time series $\{X(t): t = 0, 1, \ldots,\}$ over a finite number states $\in \mathbb{M} = \{1, 2, \ldots, m\}$. This type of forming of Markov categorical sequence was developed [20, 21].

The main advantage of using fuzzy logic is that it can handle imprecise data by assigning degrees of truth, called membership values in the closed interval [0,1]. It is an extension of the crisp method to a multivariate format. Since 1965, fuzzy set theory

[19] has occupied various fields of our society, see [17, 18, 22]. For fitting a higher order multivariate Markov chain model, basic details can be found in Ref. [23].

We now focus on the data sequence $\{X_t = j\}$. Let the initial probability vectors of size "m" in canonical form be denoted by $\mathbf{X}_t = \Pr(X_t = j) = (0, 0, \dots, 1, 0, \dots, 0)^T\}$ for t = 0,1, ... , n. Each m-vector $\mathbf{X}_t$ produces a conditional probability distribution of the categorical sequence for a given t. The corresponding normal form of the probability vector be $\overline{X}(t) = \frac{1}{n+1}\sum_{t=0}^{n}\mathbf{X}_t$.

Hence, each probability vector of size "m" of the sequence $\{\overline{X}(t) = (x_1(t), x_2(t), \dots, x_m(t))\}$ produces a conditional probability distribution of the categorical sequence for a given t. If the probability distribution at time t = r + 1 is dependent on the state probability vectors of t = r, r-1, r-2, .., r-n + 1, then the nth order model is formulated as follows:

$$\overline{\mathbf{X}}_{r+1} = \sum_{h=1}^{n} \lambda_h \, \mathbf{P}_h \, \overline{\mathbf{X}}_{r-h+1} \tag{30}$$

The initial probability vectors are $\overline{X}(t) : t = 0, 1, .., (n-1)$.$\lambda_h$ are non-negative real numbers with $\sum_{h=1}^{n}\lambda_h = 1$ and each $\mathbf{P}_h$ is the h-step transition probability matrix with each column sum equals unity for h = 1, 2, ... , n. Numerical evaluation of P_h matrices and the scalars λ_h for h = 1, 2, ... , n can be done with the "markovchain" package. Now, we can make use of the higher-order Markov model fitted by (19) for prediction future states after n-steps. The prediction of the next state $\overline{X}(t) = P_{(t-1)}\overline{X}(t\text{-}1)$ at time t can be taken as the state with the maximum probability, i.e.,

$$\overline{X}(t) = j \text{ if } \overline{X}(t) = k \le \overline{X}(t) = j \text{ k}, \text{j} = 1, 2, \dots \text{m} \tag{31}$$

To evaluate the performance and effectiveness of the HOMM, a prediction error (PE) is defined (21):

$$PE = \frac{1}{K}\sum_{j=1}^{K}\delta_j \text{ here } \quad \delta_j = \begin{cases} 1 \text{ if } \ \overline{x}_t = j \neq x_t \\ 0 \text{ if } \ \overline{x}_t = j = x_t \end{cases} \tag{32}$$

Example: Categorical data sequence and Fuzzy time series We now consider a hypothetical data on amount of rainfall (RF) time series $X(t) = \{x_t\}$ observed from t = 1991 to t = 2001 years which is reported in **Table 1**. Let the universe of discourse of the ARF factor be U = [100, 1300]. We partition the set U into 4 equal length intervals as follows: u_1 = [100, 400), u_2 = [400, 700), u_3 = [700, 1000), u_4 = [1000, 1300] which are mutually disjoint such that $U = \cup_{(j=1)}^{4} u_j$.

Let us define fuzzy sets to represent each division u_j of U through a linguistic variable, say A, as follows: A_1—low if $x(t) \in u_1$, A_2—medium if $x(t) \in u_2$, A_3—high if $x(t) \in u_3$, A_4—very high if $x(t) \in u_4$. Thus, there is a one to one correspondence between the rainfall amount x(t) received at time "t" and the values A_j for j = 1,2,3 and 4 of the linguistic variable A.

The observed categorical data sequence for the rainfall amount recorded in **Table 1** is coded by S_1 = "2," "1," "3," "2," "1," "3," "4," "4," "2," "1," "2" or by S_1 = "A_2," "A_1," "A_3," "A_2," "A_1," "A_3," "A_4," "A_4," "A_2," "A_1," "A_2." This representation given in the sequence S1 forms a fuzzy time series. In addition, the collection of all members of S_1 i.e., $\mathbb{S} = \{\text{"}A_1\text{"}, \text{"}A2\text{"}, \text{"}A3\text{"}, \text{"}A4\text{"}\}$ can be viewed as the finite state space of a Markov

Year	o(ARF)	o(S_1)	E(S_1)	E(ARF)	o(S_2)	E(ARF)
1991	500.8	A_2	NA	NA	NA	NA
1992	254.7	A_1	A_1	250	NA	NA
1993	982.3	A_3	A_3	850	A_2A_3	A_2 A_2
1994	590.9	A_2	A_2	550	A_1A_2	A_1 A_2
1995	153.4	A_1	A_1	250	A_3A_1	A_3 A_4
1996	875.4	A_3	A_3	850	A_2A_3	A_2 A_4
1997	1250.5	A_4	A_4	1150	A_1A_4	A_1 A_4
1998	1125.7	A_4	A_4	1150	A_3A_4	A_3 A_4
1999	622.6	A_2	A_2	550	$A4A_2$	$A4A_2$
2000	272	A_1	A_1	250	$A4A_1$	$A4A_1$
2001	623.4	A_2	A_3	850	$A2A_2$	A_2 A_2

A Rainfall during 1991–2001. *ARF = Rainfall amount, S1 = Linguistic Variable for ARF and expected states E(S1) from fitted Markov Model* $X_{(t+1)} = \boldsymbol{P_1} X_t$ *and* $X_{(t)} = \boldsymbol{P_2} X_{(t-1)}$.

Table 1.
HOMM model fitting.

Chain (MC) process over a period of time "t" but appropriate order of the MC is not known.

Using the sequence S_1, one can fit the first-order MC (FOMC) $\overline{X}(t+1) = \lambda_1 \mathbf{P}(t)\overline{X}(t)$ for predicting the next state from a given state of the state space $\mathbb{S} = \{\text{"}A_1\text{"}, \text{"}A_2\text{"}, \text{"}A_3\text{"}, \text{"}A_4\text{"}\}$.

By combining two adjacent states of the sequence S_1 of 11 elements, we can obtain the initial sequence S_2 = {A_2 A_1, A_1A_3, A_3A_2, A_2A_1, A_1A_3, A_3A_4, A_4A_4, A_4A_2, A_2A_1, A_1A_2} of 10 elements for constructing the second-order MC (SOMC). However, in each pair, say A_jA_k, the second position A_k is representing the current state since the Markov property forgets the past once the current state is known.

Fitting of First- and Second-order Markov models Let us fit FOMC to the observed sequence S_1 = "2," "1," "3," "2," "1," "3," "4," "4," "2," "1," "2" using the "markovchain' software package applied in R-programming. The command fitHigherOrder(S_1,order = 2) is executed and the following results for the parameters of the HOMM given in (19) are obtained:

$$\overline{\mathbf{X}}_{r+1} = \lambda_1\, \mathbf{P}_1\, \overline{\mathbf{X}}_r, \quad \text{and}, \overline{\mathbf{X}}_{r+1} = \lambda_1\, \mathbf{P}_1\, \overline{\mathbf{X}}_r + \lambda_2\, \mathbf{P}_2\, \overline{\mathbf{X}}_{(r-1)}, \quad r \geq 1 \tag{33}$$
$$\overline{\mathbf{X}}_0 = (3/11, 4/11, 2/11, 2/11), \quad \overline{\mathbf{X}}_1 = (3/10, 2/10, 3/10, 2/10)$$

λ_1 = 0.5360825 and λ_2 = 0.4639175 = 1-λ_1.
Estimates of the transition matrices:

$$\hat{P}_1 = \begin{pmatrix} 0 & 1 & 0 & 0 \\ 0.33 & 0 & 0.5 & 0.5 \\ 0.67 & 0 & 0 & 0 \\ 0 & 0 & 0.5 & 0.5 \end{pmatrix} \hat{P}_2 = \begin{pmatrix} 0 & 0 & 0.5 & 0.5 \\ 0.5 & 0.33 & 0.5 & 0.5 \\ 0 & 0.67 & 0 & 0 \\ 0.5 & 0 & 0.5 & 0 \end{pmatrix}$$

Expected States to be reached in a single step:

Consider the fit $\overline{\mathbf{X}}_1 = \lambda_1 \mathbf{P}_1 \overline{\mathbf{X}}_0$. Because the first column of the matrix $\hat{P}_1$ has the highest probability of 0.67 in the relation of the third row, which is the expected state after one step of state A.$_1$ is A_3, that is, $A_1 \rightarrow A_3$. Thus, the expected AMR is calculated as the average of the limits from u_3 = [700, 1000), i.e., E(AMR) = (700 + 1000)/ 2 = 850.

The maximum probability of the second column of the matrix $\hat{P}_1$ occurs in relation to the first row, starting from the observed state A_2, to which the MC would reach the state A_1 and the expected AMR, which is calculated from u_1 = [100 00) is 250. Our conclusion is $A_2 \rightarrow A_1$.

The maximum probability (i.e., 0.5) of the third column of the matrix $\hat{P}_1$ occurs in the second and fourth rows. Thus, starting from the considered state A_3, the MC would reach either state A_2 or A_4 and the expected AMR calculated by u_2 = [00700) is 550 and the expected AMR calculated by u_4 = [1000,1300) is 1150: P($A_3 \rightarrow A_2$) = P ($A_3 \rightarrow A_4$) = 0.5.

The maximum probability of the fourth column of the matrix $\hat{P}_1$ is 0.5 relative to rows 2 and 4. Thus, starting from the observed state A_4, the MC would reach either state A_2 or A_4, and the expected AMR calculated by u_2 = [400700) is 550 and the expected AMR calculated by u_4 = [1000,1300) is 1150: P($A_4 \rightarrow A_2$) = P($A_4 \rightarrow A_4$) = 0.5.

When comparing the states specified in columns o(ARF) and E(S_1) of the table-(\ref.*{T1}), the prediction error, or PE, is calculated using the formula (\ref) *{ER}): PE = (1/ 10) = 0.1.

Steady state stationary distribution $\overline{\mathbf{X}}$: Since the unit step transition probability matrix of the first order MC is $\hat{P}_1$, the steady state stationary distribution $\overline{\mathbf{X}}$ is the solution of

$$\overline{\mathbf{X}} = \mathbf{P}_1 \overline{\mathbf{X}} \tag{34}$$

Executing the command steadyStates($\mathbf{P}_1$) in the "markovchain" package in R code, we obtain that $\overline{\mathbf{X}} = (0.3, 0.3, 0.2, 0.2)$.

Second-order MC: Steady state stationary distribution $\overline{\mathbf{X}}^{(2)}$ Consider the initial sequence S_2 = {A_2 A_1, A_1A_3, A_3A_2, A_2A_1, A_1A_3, A_3A_4, A_4A_4, A_4A_2, A_2A_1, A_1A_2} of 10 elements for constructing the SOMC. Each state has a pair of items such as A_jA_k in which A_k is the current state. We now discuss the analysis of the model fit using the estimates of λ values and the TPMs in the following expression:

$$\overline{\mathbf{X}}_2 = \lambda_1 \mathbf{P}_1 \mathbf{X}_1 + \lambda_2 \mathbf{P}_2 \overline{\mathbf{X}}_0 \tag{35}$$

Applying the similar arguments on the factor $\overline{\mathbf{X}}_1 = \mathbf{P}_2 \overline{\mathbf{X}}_0$, we can arrive at the following conclusions: C1: $A_1 \rightarrow A_2$ or $A_1 \rightarrow A_4$.

$$A_2 \rightarrow A_3$$

$$A_3 \rightarrow A_1 \text{ or } A_3 \rightarrow A_4$$

$$A_4 \rightarrow A_1 \text{ or } A_2 \rightarrow A_2$$

Repeating the similar exercise on the factor $\overline{\mathbf{X}}_2 = \mathbf{P}_1 \overline{\mathbf{X}}_1$, we can arrive at the following conclusions:

C2

$$A_1 \rightarrow A_3$$

$$A_2 \rightarrow A_1$$

$$A_3 \rightarrow A_2 \text{ or } A_3 \rightarrow A_4$$

$$A_4 \rightarrow A_2 \text{ or } A_4 \rightarrow A_4$$

Now, ordering the results of C1 with C2 in this order we obtain the expected second-order sequences at the end of the second state: A_1 A_1 or A_1 A_2 or A_1 A_4.

$$A_2\ A_2 \text{ or } A_2\ A_4$$

$$A_3\ A_3 \text{ or } A_3\ A_2, A_3\ A_4$$

$$A_4\ A_1 \text{ or } A_4\ A_2 \text{ or } A_4\ A_4$$

All these results are reported in **Table 1**.

Prediction error for SOMC fit is PE = (3/9) = 0.33.

It can be shown that the matrix $\mathbf{P} = \lambda_1\mathbf{P}_1 + \lambda_2\mathbf{P}_2$ serves as the TPM of the SOMC. Hence, the steady state stationary distribution $\overline{\mathbf{X}}^{(2)}$ can be obtained by solving the equation $\overline{\mathbf{X}}^{(2)}\ \mathbf{P} = \overline{\mathbf{X}}^{(2)}.\overline{\mathbf{X}}^{(2)} = (0.2673632, 0.3203355, 0.1946253, 0.2176760)$.

The prediction error of the FOMC is smaller than that of second-order MC. In addition, there is no significant improvement in the four component stationary probability vector $\overline{\mathbf{X}}^{(2)}$ as compared to the respective components of first-order vector $\overline{\mathbf{X}}$. So we can conclude that the fFOMC is preferable over the SOMC for future predictions.

5. Conclusion

This chapter looked at three types of Markov models. The first model addressed the basic issues of a warehouse connected to an M/M/1/N/FCFS facility. The long-run probability distributions of queue length and inventory variables were obtained using matrix methods. The expected total cost function was formulated based on holding costs, ordering costs, and waiting time cost and was used to obtain the optimal order quantity. A graphical representation of the relationship between the cost function and order quantity was provided to support the theoretical results of this model.

The second model investigated the changes made by the arrival and departure phases of the M/M/1/N/FCFS queue and thus analyzed the simulated queue $M_s/M_s/1/N$ for long-term evaluation of moments. The input queue length and waiting times for all consecutive transitions in the $M_s/M_s/1/N$ system were calculated and all were plotted on a control chart to observe the effect of the customer waiting time distribution.

The third model analyzed the HOMM to predict the fuzzy time series as a data series. Two special cases, the FOMC and the SOMC, were fitted to the same dataset generated by measuring the hypothetical rainfall ensemble. Forecast errors for both FOMC and SOMC were calculated and compared to show that the FOMC performed best.

The best conclusions are thus made by analyzing MC models for both waiting line and inventory problems. We developed either numerically tractable expressions or accurate probabilistic expressions to obtain numerical representations. The first model can be recommended for any QIM problem and the second model to quickly assess waiting times in medical resources to ensure the best quality of service. A third model can be conveniently used to predict future states of categorical datasets with higher orders.

Future research will develop intelligent and easy-to-implement algorithms for the interdisciplinary researchers.

Conflict of interest

The author declares no conflict of interest.

Thanks declaration

The author would like to thank two colleagues for their comments on Markov models, which greatly helped this chapter of the book.

Author details

Sivasamy Ramasamy*† and Wilford B. Molefe†
Statistics Department, University of Botswana, Gaborone, Botswana

*Address all correspondence to: sivasamyr@ub.ac.bw

† These authors contributed equally.

References

[1] Gross D, Harris SM. Fundamentals of Queueing Theory. 3rd ed. New York, NY: Wiley; 1998

[2] Bhat N. An Introduction to Queueing Theory: Modeling and Analysis in Applications. 2nd ed. Boston: Birkhauser; 2008. p. 268. DOI: 10.1007/978-0-8176-4725-4

[3] Spedicato GA. Discrete Time Markov Chains with R. The R Journal. 2017. Available from: https://journal.r-project.org/archive/2017/RJ-2017-036/index.html

[4] Hunter JJ. Filtering of Markov renewal queues, II: Birth-death queues. Advanced in Application Problem. 1983; **15**:376-391

[5] Sivasamy R, Jayanthi S. Charts for detecting small shifts. Economic Quality Control. 2003;**8**(1):1-8

[6] Sivasamy R, Thillaigovindan N, Paulraj G, Paranjothi N. Quasi-birth and death processes of two server queues with stalling. OPSEARCH. 2019;**56**:739-756.. DOI: 10.1007/s12597-019-00376-1

[7] Sivasamy R. An M/M(1)+M(2)/1 queue with a state dependent K-policy. Journal of Xi'an University of Architecture & Technology. 2021;**I**: 111-124

[8] Taylor HM, Karlin S. An Introduction to Stochastic Modeling. 3rd edition. San Deigo, CA: Academic Press; 1998. ISBN-10: 0126848874, ISBN-13: 978-0126848878

[9] Zhang Y, Yue D, Sun L, Zuo J. Analysis of the queueing-inventory systems with impatient customers and mixed sales. Discrete Dynamics in Nature and Soceity. 2022;**2022**:1-12. DOI: 10.1155/2022/2333965

[10] Alnowibet KA, Alrasheedi AF, Alqahtani FS. Queueing models for analysing the steady-state distribution of stochastic inventory systems with random Lead time and impatient customers. Process. 2022;**10**:1-24. DOI: 10.3990//pr10040624

[11] Daduna H, Szekli R. M/M/1 queueing systems with inventory. Queueing System. 2006;**54**:55-78. DOI: 10.1007/s11134-006-8710-5

[12] Schwarz M, Daduna H. Queueing systems with inventory management with random lead times and with backordering. Mathematical Methods of Operational Research. 2006;**64**:383-414

[13] Latouche G, Ramaswami V. Introduction to Matrix Analytic Methods in Stochastic Modeling: ASA-SIAM Series on Statistics and Applied Probability. Philadelphia, PA: SIAM; 1999

[14] Shore H. General control charts for attributes. IIE Transactions. 2000; **32**(12):1149-1160

[15] Berman O, Kim E. Stochastic models for inventory management at service facilities. Communication Statistical Stochastic Model. 1999;**15**(4):695-718

[16] Berman O, Kim E. Dynamic order replenishment policy in internet-based supply chains. Mathematical Method Operations Research. 2001;**2001**:53

[17] Song Q, Chissom BS. Forecasting enrollments with fuzzy time series – Part I. Fuzzy Sets and Systems. 1993;**54**: 1-9

[18] Wong W-K, Bai E, Chu AW. Adaptive time-variant models for fuzzy-time-series forecasting. IEEE

Transaction System on Man Cybernetics: Part B. 2010;**40**:1531-1542

[19] Zadeh LA. Fuzzy sets. Information and Control. 1965;**8**(3):338-353. DOI: 10.1016/S0019-9958(65)90241-X

[20] Ching WK, Fung ES, Ng MK. Higher-Order Markov Chain models for categorical data sequences. Naval Research Logistics. 2004;**2004**:557-574. DOI: 10.1002/nav.20017

[21] Ching WK, Ng MK, Fung ES. Higher-order multivariate Markov chains and their applications. Linear Algebra and Its Applications. 2008;**428**: 492-506

[22] Singh P, Borah B. Forecasting stock index price based on M-factors fuzzy time series and particle swarm optimization. International Journal of Approximate Reasoning. 2014;**55**: 812-833. DOI: 10.1016/j.ijar.2013.09.014

[23] Sivasamy R, Senthamaraikannan K, Adebayo Fatai A, Karuppasamy KM. Higher order Markov chain fitting for data on rainfall and Rice production. YMER. 2021;**20**(10):156-170

Chapter 4

The Application of Markov and Semi-Markov Models in Transportation Infrastructure Management

Omar Thomas and John Sobanjo

Abstract

This chapter outlines how Markov and semi-Markov models can be used in Transportation Infrastructure Management Systems, in particular Pavement and Bridge Management Systems. The use of Markov models have been used in both Pavement and Bridge Management Systems for years. In more recent times the use of semi-Markov models have been introduced in Bridge Management Systems. Research has shown that if there is enough data available to develop semi-Markov models for transportation infrastructure, then this stochastic technique can be used to predict the future network level conditions and can be used in the development of preservation models for transportation infrastructure. The application of these techniques are not only limited to transportation infrastructure, but can also be applied in other areas.

Keywords: bridge management, pavement management, bridge deterioration, pavement deterioration, preservation model

1. Introduction

This chapter demonstrates how Markov and semi-Markov models are used in Transportation Infrastructure Management. Transportation agencies are interested in being able to model the deterioration of their infrastructure when no action is taken to neither maintain, repair nor rehabilitate that infrastructure. The Transportation agencies also want to know how to best estimate the extension in the service life of an infrastructure when improvement actions (maintenance, repair or rehabilitation) are done. The chapter uses examples containing real life data to demonstrate models for the following actions: deterioration or 'do-nothing' actions, improvement action and rehabilitation.

Markov chain has been used to model the performance of pavements in Pavement Management Systems (PMSs), such as The Arizona Department of Transportation (ADOT) Network Optimization System (NOS) [1, 2]. Wang et al. [2] outlined an approach to compute the transition probabilities of a Markov chain model using pavement performance data. Nasseri et al. [3] used Markov chain applications to investigate the crack histories of flexible pavements and to deduce the cause of rapid deterioration of surface cracks. Markov chain can also be used to model the deterioration of other infrastructures, including bridge elements, storm water pipes and wastewater pipes [4–6]. Yang et al. [7, 8] mentioned the use of semi-Markov processes for modeling the crack performance of flexible pavements, as a precursor to outlining how recurrent Markov chains can be used to model crack performance of flexible pavements. Semi-Markov processes are used in the deterioration models for other assets such as bridge elements and transformers [9–12]. Although sections 2 and 3 in this chapter uses pavement condition data to describe the application of stochastic processes for 'do-nothing' actions, the techniques can be applied to other transportation infrastructure, such as bridge elements [13].

2. Markov chain model

This section demonstrates a Markov chain model for flexible (asphalt) road pavement. The stochastic process, known as the Markov chain [14], can be described as follows: If $X_n = i$ describes a process such that the process is in state i at time n, and the process in state i has a fixed probability $P_{i,j}$ of being in state j, after a transition, then

$$P\{X_{n+1} = j|X_n = i, X_{n-1} = i_{n-1}, \dots, X_0 = i_0\} = P_{i,j} \tag{1}$$

for all states $i_0, i_1, \breve{U}i_{n-1}, i, j$ and all $n \geq 0$.

Consider the range of crack indices associated with flexible (asphalt) road pavement that have been assigned the respective condition states in the **Table 1**. The following equation, used by Wang et al. [2], can be used to generate transition probabilities for a Markov chain model:

$$p_{i,j}(a_k) = \frac{m_{i,j(a_k)}}{m_i(a_k)} \tag{2}$$

for $i,j = 10,9,8,7,6,5 \ \& \ 4$ where,

- $k = k^{th}$ rehabilitation action, in this case, the 'do-nothing' action, i.e. $k = 1$.
- $p_{i,j}(a_k)$ = transition probability from state i to j after action k is taken.

Crack Index (CRK) range	Condition state
$9.5 \leq CRK \leq 10$	10
$8.5 \leq CRK < 9.5$	9
$7.5 \leq CRK < 8.5$	8
$6.5 \leq CRK < 7.5$	7
$5.5 \leq CRK < 6.5$	6
$4.5 \leq CRK < 5.5$	5
$CRK < 4.5$	4

Table 1.
The range of crack indices and corresponding condition states.

- $m_{i,j}(a_k)$ = total number of miles of pavement for which the state prior to action k was i and the state after the action k was j.
- $m_i(a_k)$ = total number of miles of pavement for which the state prior to action k was i.

3. Semi-Markov model

This section demonstrates a Semi-Markov model for flexible (asphalt) road pavement. Consider a stochastic process having states 0,1,2, … , which is such that whenever it enters state $i, i \geq 0$, then: (1) it will enter the next state j with probability $P_{ij}, i, j, \geq 0$, and (2) given that the next state is j the sojourn time from i to j has distribution F_{ij}. For a semi-Markov process, the sojourn times may follow a specific distribution and the method of Maximum Likelihood Estimation (MLE) can be used to estimate the parameters of that distribution, such as that of a Weibull Distribution [13]. Before discussing how the semi-Markov process can be applied to model deterioration, it is beneficial to define the basic concepts of the MLE method. The Maximum Likelihood is a well-known technique in statistics used for deriving estimators [15].

3.1 Maximum likelihood estimation

If there is an identical and independently distributed (iid) sample, $X_1, \ldots, X_n$ from a population with probability density function (pdf) or probability mass function (pmf) $f\left(x|\theta_1, \breve{U}\theta_k\right)$, then the likelihood function is defined by

$$L(\theta|X) = L(\theta_1, \ldots, \theta_k|x_1, \ldots, x_n) = \Pi_{i=1}^{n} f(x_i|\theta_1, \ldots, \theta_k) \quad (3)$$

Let us assume that the sojourn time follows a Weibull Distribution. The pdf of the Weibull distribution according to Billington and Allan [16], Tobias and Trindade [17] is defined by:

$$f(t) = \frac{\beta}{\alpha}\left(\frac{t}{\alpha}\right)^{\beta-1} e^{-\left(\frac{t}{\alpha}\right)^{\beta}} \quad (4)$$

where α and β are the respective scale and shape parameters, and t represents the number of years that each unit of a mile of pavement segment takes to sojourn in one condition state before transitioning to another state. If $\eta = 1/\alpha$, then

$$f(t) = \beta\eta(\eta t)^{\beta-1} e^{-(\eta t)^{\beta}} \quad (5)$$

Using Eq. 3 [18] it follows that the likelihood is:

$$L(t_1, \ldots, t_n, \eta, \beta) = \left(\beta\eta^{\beta}\right)^{n} e^{-\eta^{\beta}}\left(t_1^{\beta} + \ldots + t_n^{\beta}\right)\Pi_{i=1}^{n} t_i^{\beta-1} \quad (6)$$

After differentiating the log likelihood and equating it to zero, the MLE of the parameters $\hat{\beta}$ and $\hat{\eta}$ are:

$$\hat{\beta} = \left[\frac{\sum_{i=1}^{n} t_i^{\hat{\beta}} \, ln\,(t_i)}{\sum_{i=1}^{n} t_i^{\hat{\beta}}} - \frac{1}{n} \sum_{i=1}^{n} ln\,(t_i) \right]^{-1} \tag{7}$$

and

$$\hat{\eta} = \left[\frac{n}{\sum_{i=1}^{n} t_i^{\hat{\beta}}} \right]^{1/\hat{\beta}} \tag{8}$$

respectively.

If there are n units of pavement in a particular condition state and k units have transitioned to a lower condition state (with complete individual sojourn times $t_1 < t_2 < \dots < t_k$) and the sojourn time for $n - k$ units of pavement are not known, then the sojourn times $T_1, \dots, T_n - k$ for the $n - k$ units of pavement should also be accounted for in the evaluation [13]. For the Weibull distribution, if it is assumed that the incomplete sojourn times of $T_1, \dots, T_{n-k}$ have been observed in addition to the complete sojourn times $t_1, \dots, t_k$, then the likelihood function can be expressed as:

$$L = \prod_{i=1}^{k} f(t_i|\theta) \cdot \Pi_{j=1}^{n-k} f\left(1 - F\left(T_j, |\theta\right)\right) \tag{9}$$

where i sums over all completed sojourn times, j sums over all incomplete sojourn times, and θ can be a vector [13].

$$L(t_1, \dots, t_n, \eta, \beta) = \left(\beta\eta^{\beta}\right)^{k} e^{-\eta^{\beta}} \left(t_1^{\beta} + \dots + t_k^{\beta}\right) \Pi_{i=1}^{k} t_i^{\beta-1} \Pi_{j=1}^{n-k} e^{-(\eta T_j)^{\beta}} \tag{10}$$

producing the MLE of the parameters $\hat{\beta}$ and $\hat{\eta}$:

$$\hat{\beta} = \left[\frac{\sum_{i=1}^{k} t_i^{\hat{\beta}} \, ln\,(t_i) + \sum_{j=1}^{n-k} T_j^{\hat{\beta}} \, ln\,(T_j)}{\sum_{i=1}^{k} t_i^{\hat{\beta}} + \sum_{j=1}^{n-k} T_j^{\hat{\beta}}} - \frac{1}{k} \sum_{i=1}^{k} ln\,(t_i) \right]^{-1} \tag{11}$$

and

$$\hat{\eta} = \left[\frac{k}{\sum_{i=1}^{k} t_i^{\hat{\beta}} + \sum^{n-k} k_{i=1} T_j^{\hat{\beta}}} \right]^{1/\hat{\beta}} \tag{12}$$

3.2 Semi-Markov Kernel

To demonstrate how the semi-Markov can be developed by knowing the sojourn times in a particular condition state before transitioning, one may consider the semi-Markov kernel in the form shown in Eq. 13. Ibe [19] defines the one-step transition probability $Q_{i,j}(t)$ of the semi-Markov process as:

$$Q_{i,j}(t) = P[X_{n+1} = j, G_n \leq t | X_n = i] \qquad t \geq 0 \tag{13}$$

where $Q_{i,j}(t)$ is the conditional probability that the process will be in state j next, given that it is in state i currently and that the waiting time in the current state i is no more than t. G_n is the time the process spends in i before transitioning to j. It also follows that:

$$Q_{i,j}(t) = p_{i,j} H_{i,j}(t) \tag{14}$$

where $p_{i,j}$ is defined as the transition probability of the embedded Markov chain, and

$$H_{i,j}(t) = P[G_n \leq t | X_n = i, X_{n+1} = j] \tag{15}$$

3.2.1 Semi-Markov process

Howard [20] provided the following formulation to determine the probability that a continuous-time semi-Markov process will be in state j at time n given that it entered state i at time zero.

$$\phi_{ij}(n) = \delta_{ij}\, {}^{>}w_i(n) + \sum_{k=1}^{N} p_{ik} \int_0^n h_{ik}(m) \phi_{kj}(n-m) i = 1, 2, \dots, N; j = 1, 2, \dots, N; \quad n = 0,1,2, \dots \tag{16}$$

$$\delta_{ij} = \begin{cases} 1 & i = j; \\ 0 & i \neq j. \end{cases} \tag{17}$$

where $\phi_{ij}(n)$ is the probability that a continuous-time semi-Markov process will be in state j at time n given that it entered state i at time $n = 0$ and is referred to as the interval transition probability from state i to state j in the interval $(0, n)$. ${}^{>}w_i(n)$ is the probability that the process will exit its starting state i at a time greater than n.

The second term in 3.2.1 describes the probability of the sequence of events, such that the process makes an initial transition from state i to some state k at some time m and thereafter proceeds from state k to state j in the remaining time n–m. To account for all possible scenarios, the probability is summed over all states k to which the first transition could have been made and over all times of the initial transition, m, between l and n. p_{ik} is the probability of transitioning from i to k, and $h_{ik}(m)$ represents the probability distribution of the sojourn time from i to k at time m. The matrix formulation of 3.2.1 can be expressed as:

$$\Phi(n) = {}^{>}W(n) + \int_0^n [P \square H(m)] \Phi(n-m) \qquad n = 0,1,2, \dots \tag{18}$$

In addition, let

$$C(m) = [P \square H(m)] \tag{19}$$

where C(m) is defined as the core matrix [20]. The elements of $C(m)$ are $c_{ij}(m) = p_{ij} h_{ij}(m)$, where p_{ij} represents the transition probability of the embedded Markov chain and $h_{ij}(m)$ represents the probability distribution of the sojourn time in state i, before transitioning to j at time m.

As a result, the interval transition matrix representing a single transition at time m, $\Phi^{0,m}(m)$, for 'do-nothing' actions can be expressed as:

$$\Phi^{0,m}(m) = \begin{pmatrix} 1-\sum_{j=4}^{9} p_{10,j}H_{10,j}(m) & \cdots & \cdots & \cdots & \cdots & p_{10,5}H_{10,5}(m) & p_{10,4}H_{10,4}(m) \\ 0 & \ddots & \cdots & \cdots & \cdots & p_{9,5}H_{9,5}(m) & p_{9,4}H_{9,4}(m) \\ 0 & 0 & \ddots & \cdots & \cdots & p_{8,5}H_{8,5}(m) & p_{8,4}H_{8,4}(m) \\ 0 & 0 & 0 & \ddots & \cdots & p_{7,5}H_{7,5}(m) & p_{7,4}H_{7,4}(m) \\ 0 & 0 & 0 & 0 & \ddots & p_{6,5}H_{6,5}(m) & p_{6,4}H_{6,4}(m) \\ 0 & 0 & 0 & 0 & 0 & 1-p_{5,4}H_{5,4}(m) & p_{5,4}H_{5,4}(m) \\ 0 & 0 & 0 & 0 & 0 & 0 & 1 \end{pmatrix} \tag{20}$$

where p_{ij} represents the probability for the embedded Markov chain of the semi-Markov process, and H_{ij} represents the cumulative distribution of the sojourn time between condition state i and condition state j at time m.

It can be seen by 3.2.1 that as the number of years, m, increases, a number of permutations must be considered when computing the overall transition probabilities for an interval $(0, n)$, and so another approach is suggested to model the overall transition over time. In the context of modeling asset deterioration, the focus is on 'do-nothing' scenarios, where there is no action taken on the infrastructure. It therefore can be assumed that once the infrastructure exits a condition state, that condition state is not visited again, and the semi-Markov process occurs in one direction only. There is also the probability of the condition of some infrastructure, in this case the pavement segments, 'skipping' condition states in a single transition, however the probability of 'skipping' two (2) or more condition states is extremely small and therefore may not have to be considered. If it is assumed that only one condition state may be 'skipped' in a single transition, then the transition probability (for the embedded Markov chain), p_{ij}, of 'skipping' a condition state, k, is one minus the transition probability (for the embedded Markov chain) of transitioning to the next state, p_{ik}. Therefore,

$$p_{ij} = 1 - p_{ik} \quad i = 10,9,8,7,6,5;\ j = i-2,\ k = i-1,\ min\,(j,k) = 4 \tag{21}$$

This is because the total probability of *eventually* leaving a particular condition state (which is not a terminal state), to a lower condition state is 1, since deterioration is being considered.

Another assumption is that the sojourn time in condition state i before transitioning to j is the time from the pavement segment first entered condition state i to the time the pavement segment first entered condition state j. The transition diagram shown in **Figure 1** shows the possible 'single step' transitions in developing

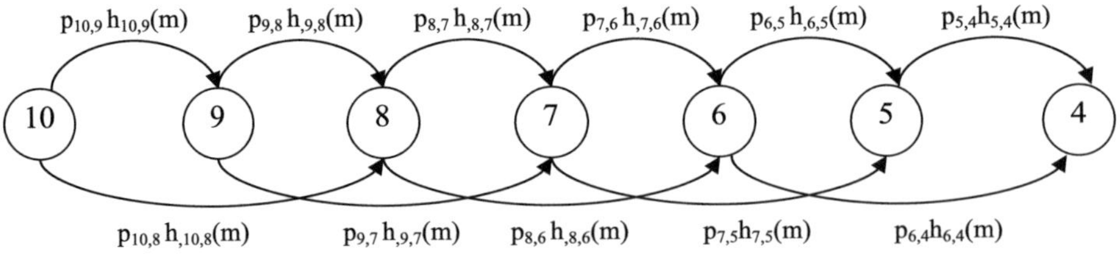

Figure 1.
Transition diagram that represents the possible 'single step' transitions between condition states.

the semi-Markov model, where p_{ij} is the probability for the embedded Markov chain from condition state i to condition state j, and $h_{ij}(m)$ is the probability density function of the sojourn time between condition state i and condition state j at time m. The term 'single step' transition is based on the assumption that only a single transition can take place in a year. For the transition labeled $p_{10,9}h_{10,9}(m)$ the pavement segment spends some time in condition state 10 before transitioning to condition state 9, and for the transition labeled $p_{10,8}h_{10,8}(m)$ the pavement segment spends some time in condition state 10 before transitioning to condition state 8 without going to 9 [13].

Other than determining the interval transition probabilities for the interval $(0, n)$, the conditional transition probabilities for each yearly interval for $m = 1, 2, \dots, n$ is determined, which is multiplied by each other to estimate the transition probabilities for the interval $(0, n)$. It is also assumed that only a single transition takes place in a year. In other words, let

$$\Phi^{0,n}(n) = \Phi^{0,1} \cdot \Phi^{1,2} \cdot \dots \Phi^{n-1,n} \tag{22}$$

where $\Phi^{m-1,m}$ is a 1 year 'single' transition probability matrix for the time $m - 1$ to m (i.e. m^{th} interval), $m = 1, 2, \dots, n$. Based on Eq. (20), if the condition states can drop by either one (1) or two (2) states, then the interval transition probability for the first year results to:

$$\Phi(m) = \begin{pmatrix} 1 - \sum_{j=8}^{9} p_{10,j}H_{10,j}(m) & p_{10,9}H_{10,9}(m) & p_{10,8}H_{10,8}(m) & 0 & 0 & 0 & 0 \\ 0 & \ddots & \dots & \dots & 0 & 0 & 0 \\ 0 & 0 & \ddots & \dots & \dots & 0 & 0 \\ 0 & 0 & 0 & \ddots & \dots & \dots & 0 \\ 0 & 0 & 0 & 0 & \ddots & \dots & \dots \\ 0 & 0 & 0 & 0 & 0 & \ddots & \dots \\ 0 & 0 & 0 & 0 & 0 & 0 & 1 \end{pmatrix} \tag{23}$$

where $m = 1$. On the other hand, to determine the respective transition probability matrices for the subsequent intervals, a different formulation is used, in which it is assumed that the sojourn time is left truncated at the start of each interval.

3.2.2 Service life analysis and left truncation

Statistical analyses of the service lives of transportation infrastructures can be done using Reliability Theory otherwise called Survival Analysis. In Survival Analysis, sometimes subjects are selected and followed prospectively until the event or censoring occurs, but sometimes the start time at the point of selection is not $t = 0$ (i.e. not at birth), but at a value $t = t_0 > 0$. It therefore means that the life or censoring time of the subjects, T_i, is greater than t_0 [21, 22], and the life T_i is considered to be left truncated at t_0. Applying the same principle to the service life of a transportation infrastructure then,

$$F_{T|T>t_0}(t) = \begin{cases} 0 & \text{if } t_0 \geq t; \\ \dfrac{F_T(t) - F_T(t_0)}{1 - F_T(t_0)} & \text{if } t_0 < t. \end{cases} \tag{24}$$

To determine the probabilities that are associated with the sojourn time for interval $(1, 2]$, we assume that from Eq. (24) $t_0 = 1$ and $1 < t \leq 2$. It therefore means that at this point only sojourn times greater than $t = 1$ are being considered and the cumulative distribution of the sojourn time in the interval can be considered truncated [22]. It is therefore can be described as:

$$H_{i,j_{T|T>1}}(t) = \frac{H_{i,j_T}(t) - H_{i,j_T}(1)}{1 - H_{i,j_T}(1)}, \qquad 1 < t \leq 2 \tag{25}$$

It follows that for interval $(m-1, m]$ the cumulative distribution of the sojourn time in the interval can be described as:

$$H_{i,j_{T|T>m-1}}(t) = \frac{H_{i,j_T}(t) - H_{i,j_T}(m-1)}{1 - H_{i,j_T}(m-1)}, \qquad m-1 < t \leq m \tag{26}$$

At $t = m$ the cumulative distribution of the sojourn time then becomes

$$H_{i,j_{T|T>m-1}}(m) = \frac{H_{i,j_T}(m) - H_{i,j_T}(m-1)}{1 - H_{i,j_T}(m-1)}, \qquad m-1 < t \leq m \tag{27}$$

Therefore, the transition probability for interval $(m-1, m]$ is:

$$\Phi^{m-1,m}(m) = \begin{pmatrix} 1 - \sum_{j=8}^{9} p_{10,j} H_{10,j_{T|T>m-1}}(m) & p_{10,9} H_{10,9_{T|T>m-1}}(m) & \cdots & 0 & 0 & 0 & 0 \\ 0 & \ddots & \cdots & \cdots & 0 & 0 & 0 \\ 0 & 0 & \ddots & \cdots & \cdots & 0 & 0 \\ 0 & 0 & 0 & \ddots & \cdots & \cdots & 0 \\ 0 & 0 & 0 & 0 & \ddots & \cdots & \cdots \\ 0 & 0 & 0 & 0 & 0 & \ddots & \cdots \\ 0 & 0 & 0 & 0 & 0 & 0 & 1 \end{pmatrix} \tag{28}$$

Eq. (27) have been used to determine the transition probabilities for one-step state-based yearly transitions in deterioration models by Black et al. [11, 12], for which the transition probability of the embedded Markov chain was assumed to be 1. It therefore means that the probability obtained using Eq. (27) can be used to describe the probability associated with the sojourn time to the end of the period, given that it had 'survived' up to the start of the period.

3.2.3 Sojourn times of flexible pavement condition states

For developing the semi-Markov model, the number of miles for a particular segment can be rounded to the nearest one-tenth of a mile and the sojourn time distribution for each one-tenth unit of a mile of pavement segment is then analyzed. **Figure 2** gives a schematic of how the pavement segment can be divided into one-tenth of a mile sub-sections. An algorithm can be created and used to help organize and analyze the data on the infrastructure to estimate the parameters of the sojourn time distributions in each condition state. The following outlines the steps:

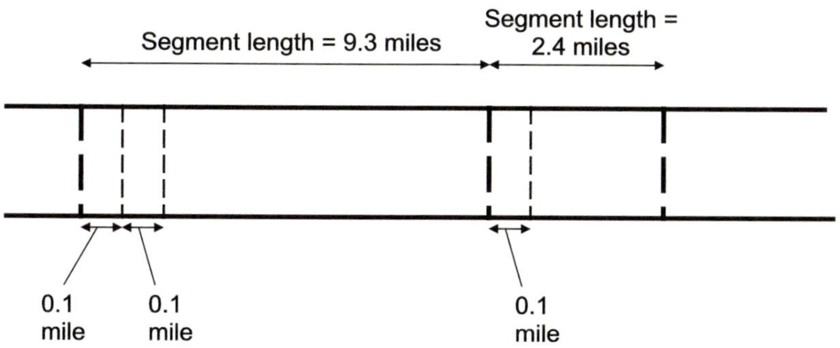

Figure 2.
Schematic of pavement segment divisions into equal units of one-tenth of a mile.

1. The length of each segment to the nearest one-tenth of a mile is determined.

2. The yearly decreases in *CRKs* for each segment subsequent to the year being newly constructed or overlaid is extracted, such that a series of decreasing *CRK*s for each segment can be used to model 'do-nothing' actions on the pavement segments over time.

3. The pavement segments are assigned to condition states over time, in accordance with **Table 1**, as outlined earlier.

4. At some point in time, all the pavement segments tracked, are either just becoming 'new' or entered the current condition state from a higher or equal condition state. If a unit of pavement segment exits a particular condition state to a lower condition state at a known time, then the sojourn time of that unit of pavement segment in the current condition state is essentially known. However, if a unit of pavement segment is in a particular condition state and the tracking of the pavement ended because the condition state either increased or the because the 'study' was terminated, then the sojourn time of that unit of pavement segment is not precisely known and is considered right-censored [23]. **Figure 3** gives a representation of the change in the condition states of pavement segments over time, outlining examples of complete and censored times spent in particular condition states.

5. Based on the complete and censored durations obtained from the data, the distribution of the sojourn time in condition state i before transitioning to condition state j ($H_{i,j}(t)$) can be determined, where j is either $i-1$ or $i-2$, $i = 10,9,8,7,6,5 \;\; min\,(j) = 4$. The proportion of the number of units of the

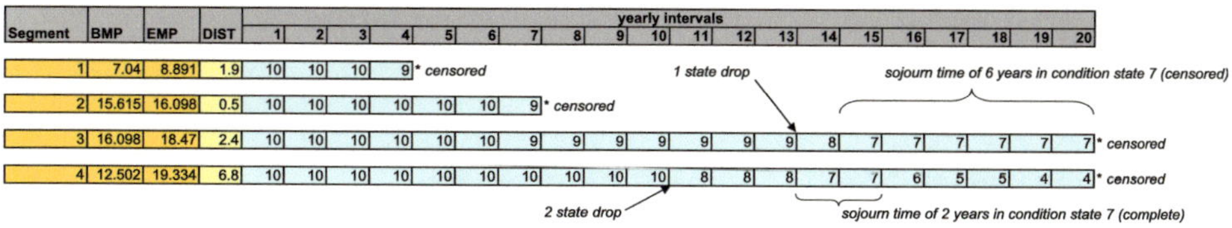

Figure 3.
An example of the change in the condition states of pavement segments over time.

infrastructure that left condition state i and transitioned to condition state j, to the total number of units of assets that left condition state i and transitioned to all condition states other than itself ($p_{i,j}$) can be determined, for each condition state i. From **Figure 3**, one can see that segment 3 spends 6 years in condition state 10, has a one state drop and then spends 7 years in condition state 9, and then has another one state drop to condition state 8. It shows that segment 3 then spends 1 year in condition state 8, before transitioning to condition state 7, where the pavement segment spent at least 6 years in condition state 7. In **Figure 3**, for segment 4, it is seen that the segment spends 10 years in condition state 10 before having a two-state drop to condition state 8, followed by a series of one state drops for which the respective sojourn times can also be inferred.

The maximum likelihood estimate of the scale (α) and shape (β) parameters of the Weibull distribution, used to describe the sojourn time distributions, can be computed. An algorithm can be written to determine the series of transition probability matrices, based on the Weibull parameters (α and β values) obtained for the sojourn times between condition states. The MLE of the α and β values serves as inputs for this algorithm. The transition probabilities according to Eqs. (23) and (28) can then be determined and used to simulate the survival curves and the expected deterioration over time.

4. Examples of transition probabilities associated with the Markov chain and semi-Markov models

4.1 Transition probabilities in a Markov chain model

An example of a set of transition probabilities for a Markov chain model are represented as a transition probability matrix, as shown below in Eq. (29). At the left and top of the transition probability matrix in Eq. (29) are the condition states i and j respectively, where Π_{ij} is transition probability matrix used for a Markov chain model.

$$\Pi_{ij} = \begin{matrix} 10 \\ 9 \\ 8 \\ 7 \\ 6 \\ 5 \\ 4 \end{matrix} \overset{\begin{matrix} 10 & 9 & 8 & 7 & 6 & 5 & 4 \end{matrix}}{\begin{pmatrix} 0.905 & 0.072 & 0.017 & 0.006 & 0 & 0 & 0 \\ 0 & 0.737 & 0.157 & 0.090 & 0.016 & 0 & 0 \\ 0 & 0 & 0.660 & 0.274 & 0.042 & 0.014 & 0.010 \\ 0 & 0 & 0 & 0.707 & 0.188 & 0.086 & 0.019 \\ 0 & 0 & 0 & 0 & 0.724 & 0.112 & 0.164 \\ 0 & 0 & 0 & 0 & 0 & 0.582 & 0.418 \\ 0 & 0 & 0 & 0 & 0 & 0 & 1 \end{pmatrix}}. \tag{29}$$

4.2 Transition probabilities in a semi-Markov model

An example of the transition probability of the embedded Markov chain of a semi-Markov process are shown for each transition in **Table 2**.

Transitions *i* to *j*	Transition probability
10 to 9	0.707
9 to 8	0.752
8 to 7	0.645
7 to 6	0.468
6 to 5	0.214
5 to 4	1.000
10 to 8	0.293
9 to 7	0.248
8 to 6	0.355
7 to 5	0.532
6 to 4	0.786

Table 2.
Transition probabilities of the embedded Markov chain of the semi-Markov process.

Figures 4–7 show examples of the frequency distributions of the observed sojourn time for each unit mile before transition, including both the uncensored and right-censored sojourn times.

From **Figures 4** and **5** it can be seen that there are more units of pavement that transitioned completely from condition state 10 to 9, than from 10 to 8, which is expected. The total pavement length that are right-censored in **Figures 4** and **5** are the same. This is because it is not known whether each unit of pavement would have transitioned from condition state 10 to 9 or from 10 to 8.

Based on the Maximum Likelihood Estimation (MLE), examples of the Weibull distribution parameters are shown in **Table 3**, while the associated uncertainties are

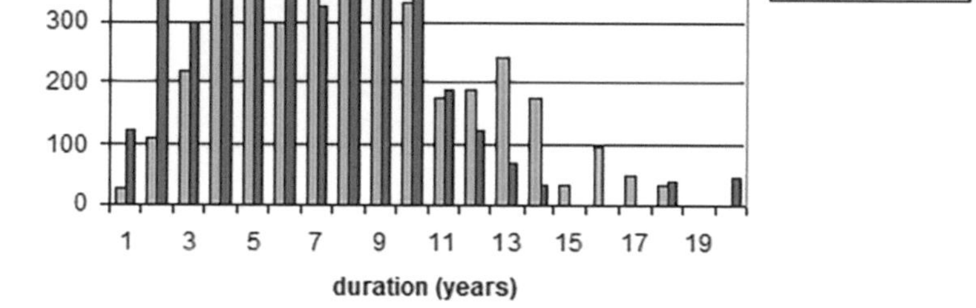

Figure 4.
Frequency of sojourn times in condition state 10 before transitioning to 9.

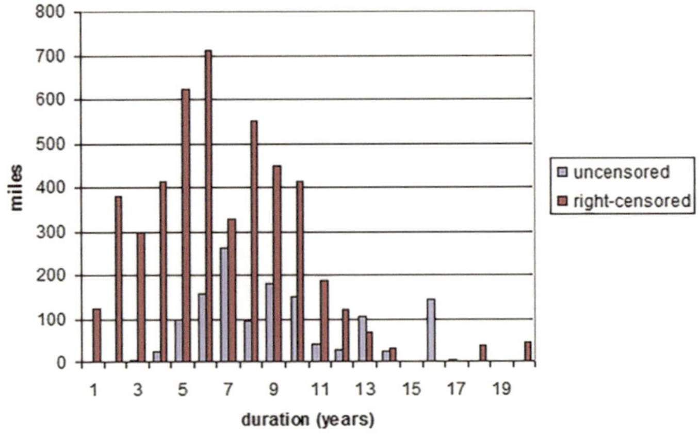

Figure 5.
Frequency of sojourn times in condition state 10 before transitioning to 8.

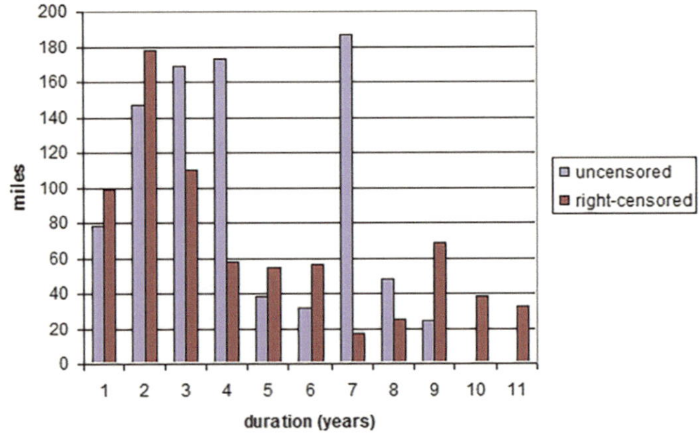

Figure 6.
Frequency of sojourn times in condition state 9 before transitioning to 8.

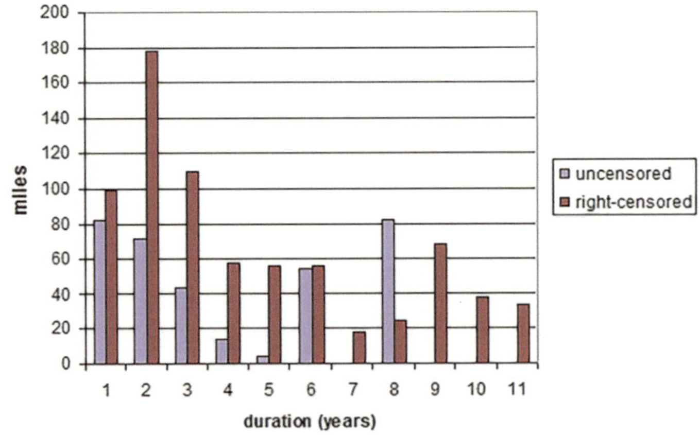

Figure 7.
Frequency of sojourn times in condition state 9 before transitioning to 7.

Transition *i* to *j*	$\hat{\alpha}$	95% C.I. limits for $\hat{\alpha}$		$\hat{\beta}$	95% C.I. limits for $\hat{\beta}$	
		Lower	Upper		Lower	Upper
10 to 9	9.432	9.332	9.533	2.128	2.094	2.163
9 to 8	4.887	4.777	4.999	1.579	1.539	1.62
8 to 7	3.496	3.394	3.602	1.345	1.304	1.387
7 to 6	5.039	4.811	5.278	1.257	1.208	1.308
6 to 5	6.304	5.754	6.906	1.523	1.412	1.641
5 to 4	3.164	3.03	3.304	2.062	1.94	2.193
10 to 8	13.126	12.904	13.351	3.182	3.088	3.278
9 to 7	6.103	5.845	6.372	1.249	1.204	1.295
8 to 6	9.672	8.744	10.697	1.465	1.35	1.591
7 to 5	9.103	8.355	9.918	1.236	1.165	1.312
6 to 4	5.417	5.092	5.763	1.693	1.591	1.802

Table 3.
Example of the maximum likelihood estimation of the scale (α) and shape (β) parameters for the holding time (Weibull) distributions in condition state i.

Transitions *i* to *j*	Mean (years)	Standard deviation
10 to 9	8.35	4.13
9 to 8	4.39	2.84
8 to 7	3.21	2.41
7 to 6	4.69	3.75
6 to 5	5.68	3.80
5 to 4	2.80	1.43
10 to 8	11.75	4.05
9 to 7	5.69	4.58
8 to 6	8.76	6.08
7 to 5	8.50	6.91
6 to 4	4.84	2.94

Table 4.
Means and standard deviations of sojourn times (Weibull).

illustrated in **Table 4**. The 95% C.I. limits of the scale and shape parameters are shown in **Table 3**. **Table 4** provides the means and standard deviations for each sojourn time (Weibull) distribution. The standard deviations for the sojourn time in condition state 8 before transitioning to 6, and that of condition state 7 before transitioning to 5 seems relatively high in comparison to the others, which is a function of the quantity of data representing each transition. As more data for each transition becomes available, then the standard deviations associated with the particular transition is expected to be less. A Goodness-of-fit test can also be done on the distribution of the complete sojourn times.

As an example, the transition probability matrices of the first 5 years for the semi-Markov model are shown in Eqs. (30)–(34). In Eq. (30), $\phi_{10,10} = 0.991$ means that there is a 0.991 probability that the pavement segment remains in condition state 10. The other $\phi_{i,j}$ terms represent the probability of transition from i to j in a given year.

Year 1

$$\Phi_{ij}(1) = \begin{array}{c} \\ 10 \\ 9 \\ 8 \\ 7 \\ 6 \\ 5 \\ 4 \end{array} \begin{array}{c} \begin{array}{ccccccc} 10 & 9 & 8 & 7 & 6 & 5 & 4 \end{array} \\ \begin{pmatrix} 0.991 & 0.008 & 0.001 & 0 & 0 & 0 & 0 \\ 0 & 0.822 & 0.078 & 0.100 & 0 & 0 & 0 \\ 0 & 0 & 0.795 & 0.170 & 0.035 & 0 & 0 \\ 0 & 0 & 0 & 0.814 & 0.123 & 0.063 & 0 \\ 0 & 0 & 0 & 0 & 0.886 & 0.059 & 0.056 \\ 0 & 0 & 0 & 0 & 0 & 0.911 & 0.089 \\ 0 & 0 & 0 & 0 & 0 & 0 & 1 \end{pmatrix} \end{array} . \tag{30}$$

Year 2

$$\Phi_{ij}(2) = \begin{array}{c} \\ 10 \\ 9 \\ 8 \\ 7 \\ 6 \\ 5 \\ 4 \end{array} \begin{array}{c} \begin{array}{ccccccc} 10 & 9 & 8 & 7 & 6 & 5 & 4 \end{array} \\ \begin{pmatrix} 0.970 & 0.028 & 0.002 & 0 & 0 & 0 & 0 \\ 0 & 0.716 & 0.150 & 0.134 & 0 & 0 & 0 \\ 0 & 0 & 0.690 & 0.249 & 0.061 & 0 & 0 \\ 0 & 0 & 0 & 0.749 & 0.166 & 0.085 & 0 \\ 0 & 0 & 0 & 0 & 0.773 & 0.107 & 0.120 \\ 0 & 0 & 0 & 0 & 0 & 0.744 & 0.256 \\ 0 & 0 & 0 & 0 & 0 & 0 & 1 \end{pmatrix} \end{array} . \tag{31}$$

Year 3

$$\Phi_{ij}(3) = \begin{array}{c} \\ 10 \\ 9 \\ 8 \\ 7 \\ 6 \\ 5 \\ 4 \end{array} \begin{array}{c} \begin{array}{ccccccc} 10 & 9 & 8 & 7 & 6 & 5 & 4 \end{array} \\ \begin{pmatrix} 0.944 & 0.049 & 0.007 & 0 & 0 & 0 & 0 \\ 0 & 0.652 & 0.197 & 0.151 & 0 & 0 & 0 \\ 0 & 0 & 0.633 & 0.290 & 0.077 & 0 & 0 \\ 0 & 0 & 0 & 0.717 & 0.188 & 0.095 & 0 \\ 0 & 0 & 0 & 0 & 0.695 & 0.138 & 0.167 \\ 0 & 0 & 0 & 0 & 0 & 0.602 & 0.398 \\ 0 & 0 & 0 & 0 & 0 & 0 & 1 \end{pmatrix} \end{array} . \tag{32}$$

Year 4

$$\Phi_{ij}(4) = \begin{matrix} \\ 10 \\ 9 \\ 8 \\ 7 \\ 6 \\ 5 \\ 4 \end{matrix} \begin{matrix} 10 & 9 & 8 & 7 & 6 & 5 & 4 \\ \end{matrix} \begin{pmatrix} 0.915 & 0.071 & 0.014 & 0 & 0 & 0 & 0 \\ 0 & 0.603 & 0.234 & 0.163 & 0 & 0 & 0 \\ 0 & 0 & 0.591 & 0.319 & 0.090 & 0 & 0 \\ 0 & 0 & 0 & 0.694 & 0.203 & 0.103 & 0 \\ 0 & 0 & 0 & 0 & 0.631 & 0.163 & 0.206 \\ 0 & 0 & 0 & 0 & 0 & 0.484 & 0.516 \\ 0 & 0 & 0 & 0 & 0 & 0 & 1 \end{pmatrix}. \quad (33)$$

Year 5

$$\Phi_{ij}(5) = \begin{matrix} \\ 10 \\ 9 \\ 8 \\ 7 \\ 6 \\ 5 \\ 4 \end{matrix} \begin{matrix} 10 & 9 & 8 & 7 & 6 & 5 & 4 \\ \end{matrix} \begin{pmatrix} 0.883 & 0.093 & 0.024 & 0 & 0 & 0 & 0 \\ 0 & 0.562 & 0.265 & 0.173 & 0 & 0 & 0 \\ 0 & 0 & 0.557 & 0.343 & 0.100 & 0 & 0 \\ 0 & 0 & 0 & 0.676 & 0.215 & 0.109 & 0 \\ 0 & 0 & 0 & 0 & 0.577 & 0.183 & 0.240 \\ 0 & 0 & 0 & 0 & 0 & 0.388 & 0.612 \\ 0 & 0 & 0 & 0 & 0 & 0 & 1 \end{pmatrix}. \quad (34)$$

5. Stochastic preservation model

This section outlines a stochastic preservation model based on bridge elements. The model can be used for transportation infrastructure management, in which the underlining decision model is based on a semi-Markov decision process (SMDP). The "Bare Concrete Deck" is an example of a bridge element that has different condition states to which the preservation model can be applied in a Bridge Management System. The preservation model is adaptive and this concept has been used by researchers in the field of infrastructure management, particularly as it relates to pavement and rail infrastructures [24–27]. Madanat et al. [27] also describes the methodology used in the Pontis® Bridge Management System as being adaptive, since transition probabilities are updated over time. The methodology outlined here is based on the application of SMDP.

The maintenance activities on a bridge element can encompass a wide range of types of work having varying durations. Again, the Weibull distribution can be assumed to estimate the sojourn time, which in this case, is the time it takes for maintenance actions to be completed. Unlike the 'do-nothing' action, it is assumed that the sojourn times for maintenance activities are irrespective of the current and subsequent condition states of the bridge element. It can also be assumed that rehabilitation works predominantly encompass the replacement of the respective bridge element, which has a more predictable duration time than that of maintenance works. The time taken for improvement (maintenance and rehabilitation) works to be completed, can be interpreted to start from the point at which the problem to the bridge element was first identified, including the time for the required works to be sent out for bid if necessary, followed by the time taken to undertake the works.

5.1 Network level optimization

One of the main goals of a Bridge Management System (BMS) is to be able to determine the minimum-cost long-term policy for each bridge element [28]. This policy consists of a set of recommended actions that minimizes the long-term Maintenance, Repair and Rehabilitation (MR&R) cost requirements, while keeping the bridge element out of risk of failure. If the minimum-cost long-term policy can be determined, it represents the most cost-efficient set of actions for the bridge element. Therefore, if any of the actions are delayed, it results in more expense in the long-term. Also, if more improvement actions other than what is recommended are done, then it will also result in higher costs in the long-term. This is based on the steady state concept. Bridge elements are expected to remain in service for long periods of time providing transportation connectivity on a continuous basis. Having an optimal policy that is sustainable far into the future is of great importance. In a BMS the following three (3) things typically takes place each year: (1) Bridge elements deteriorate when no improvement actions are done to them, otherwise called 'do-nothing' actions; (2) Improvement actions are done to some of the bridge elements, which results in corresponding costs; and (3) The improvement actions cause an overall improvement in network conditions [13].

At the network level, any given condition state will have elements passing in and out of condition states based on the MR&R actions. For steady state to occur across the entire bridge network or across a subset of that network, the total quantity of bridge element entering a particular condition state is equal to the total quantity of that similar bridge element leaving that same condition state. As a result the distribution of a bridge element among its condition state remains constant from year to year within the group, and the policy becomes sustainable over the long-term [28]. The optimal policy is the one policy that satisfies the requirements of a steady state and consequently minimizes annual expenditure of the transportation agency in charge of MR&R works.

The Markov Decision Process MDP have been used in the Pontis Bridge Management System to determine the optimal policy, in which there is a means to update the transition probabilities over time or as needed. This is because a Markov chain model assumes that the sojourn time in one state before transitioning to another follows an exponential distribution for continuous time, thus making it more restrictive than the semi-Markov process in which the sojourn times can be assumed to follow a different distribution, such as the Weibull distribution. The SMDP facilitates the natural updating or changing of the transition probabilities over time. Again, it is assumed that the sojourn time in a particular condition state follows the Weibull distribution, and the deterioration process can be represented as a semi-Markov process. The probability density function (pdf) for the Weibull distribution is [29]:

$$f(t) = \frac{\beta}{\alpha}\left(\frac{t}{\alpha}\right)^{\beta-1} e^{-\left(\frac{t}{\alpha}\right)^{\beta}} \tag{35}$$

where α and are the scale and shape parameters respectively. When the shape parameter equals 1, the resulting distribution becomes exponential, and the rate deterioration is constant. If the shape parameter is greater than 1, it means that the rate of deterioration increases with time, and if the shape parameter is less than 1, the rate of deterioration decreases with time. Typically, the latter is not expected for transportation infrastructure.

One component of the preservation model for bridge elements is the deterioration model similar to that explained above, where the Maximum Likelihood Estimation of the parameters of the Weibull distribution, used to describe the sojourn time in one condition state before transitioning to a lower condition state, is determined. A similar approach can be used to determine the sojourn time distribution for maintenance works.

5.2 Discount coefficient

Let us consider continuous-time discounting at a rate $s > 0$, such that the present value of one unit received t times units in the future equals e^{-st}. For discounting over 1 year, let $t = 1$, therefore, $e^{-s} = d$, where d represents corresponding discrete-time discount rate that would be used in a MDP. So, for example, $d = 0.9$ in the MDP corresponds to $s = -\log(0.9) = 0.105$ in the SMDP, which is the corresponding discount factor based on continuous-time [30].

5.3 The Laplace transform (s-transform)

When taking into consideration the discounting for continuous-time in the SMDP model, the Laplace transform (s-transform) of the distribution of the sojourn times between states has to be computed [19]. It is therefore prudent to look at Laplace transform. Now, if is the pdf of a continuous random variable X that takes only non-negative values; then $f_X(x) = 0$ for $x < 0$. It follows that the Laplace transform of $f_X(x)$, denoted by $M_x(s)$ is [19]:

$$M_x(s) = E\left[e^{-sX}\right] = \int_0^{\infty} e^{-sx} f_x(x)dx \tag{36}$$

An essential Laplace transform property is that when it is evaluated at s = 0, its value is equal to 1:

$$M_x(s)|_{s=0} = \int_0^{\infty} f_x(x)dx = 1 \tag{37}$$

The Laplace transform of the Weibull distribution can be complex, as shown in Eq. (38) [31].

$$E\left[e^{-tX}\right] = \frac{1}{\lambda^k t^k} \frac{p^k \sqrt{q/p}}{\left(\sqrt{2\pi}\right)^{q+p-2}} G_{p,q}^{q,p}\left(\begin{matrix} \frac{1-k}{p}, \frac{2-k}{p}, \ldots, \frac{p-k}{p} \\ \frac{0}{q}, \frac{1}{q}, \ldots, \frac{q-1}{q} \end{matrix} \middle| \frac{p^p}{\left(q\lambda^k t^k\right)^q} \right) \tag{38}$$

where G is known as the Meijer G-function.

As a result, it may have to be solved numerically. The numerical solutions are determined by substituting the scale and shape (Weibull) parameter estimates and the continuous-time discount rate, s, into the expression provided in Eq. (39).

$$M_x(s) = E\left[e^{-sX}\right] = \int_0^{\infty} e^{-sx} \cdot \frac{\beta}{\alpha}\left(\frac{t}{\alpha}\right)^{\beta-1} e^{-\left(\frac{t}{\alpha}\right)^{\beta}} dx \tag{39}$$

For the case of when the bridge element condition is in the terminal state, mathematically, it can be assumed that the bridge element spends 1 year in that state before 'transitioning' back into the terminal state. When a particular transition takes 1 year every time, then the equivalent to the Laplace transform in this scenario is provided by Eq. (40) [30]:

$$M_x(s) = E\left[e^{-sX}\right] = e^{-s} \tag{40}$$

5.4 Semi-Markov decision process with discounting

The computation of the present values based on the SMDP is provided by the following formulation [19, 20]:

$$v_i(a,\lambda) = r_i(a,\lambda) + \sum_{j=1}^{N} p_{ij}(a) \int_{\tau=0}^{\infty} v_j(a,\lambda) e^{-\lambda\tau} f_{H_{ij}}(\tau, a) d\tau \qquad i = 1, \dots, N \tag{41}$$

where, $v_i(a,\lambda)$ = total expected long-term discounted cost; i = the condition state of a bridge element; a = set of feasible MR&R actions for condition state i; $r_i(a,\lambda)$ = the expected cost for an MR&R action a in condition state i; λ = the continuous-time discount factor; $p_{ij}(a)$ = the transition probability of the embedded Markov chain that the bridge element will transition from condition state i to condition state j after action a is taken; $f_{H_{ij}}$ is the probability density function of the sojourn time, τ, from condition state i to condition state j, when action a is taken. Therefore,

$$v_i(a,\lambda) = r_i(a,\lambda) + \sum_{j=1}^{N} p_{ij}(a) M_{ij}^{H}(a,\lambda) v_j(a,\lambda) \tag{42}$$

where $M_{ij}^{H}(a,\lambda)$ is the Laplace transform (s-transform) of $f_{H_{ij}}(\tau, a)$ evaluated at $s = \lambda$, and transition probability matrix, $P_{i,j}$ is given by,

$$P_{i,j}(a) = \begin{bmatrix} p_{1,1}(a) & p_{1,2}(a) & \dots & \dots & p_{1,N}(a) \\ p_{2,1}(a) & p_{2,2}(a) & \dots & \dots & p_{2,N}(a) \\ \dots & \dots & \dots & \dots & \dots \\ p_{N-1,1}(a) & p_{N-1,2}(a) & \dots & \dots & p_{N-1,N}(a) \\ p_{N,1}(a) & p_{N,2}(a) & \dots & \dots & p_{N,N}(a) \end{bmatrix} \tag{43}$$

If we let

$$q_{ij}(a,\lambda) = p_{ij}(a) M_{ij}^{H}(a,\lambda) \tag{44}$$

we obtain

$$v_i(a,\lambda) = r_i(a,\lambda) + \sum_{j=1}^{N} q_i(a,\lambda) v_j(a,\lambda) \qquad i = 1, \dots, N. \tag{45}$$

According to [19], the long-term value for $r_i(a,\lambda)$ is given by

$$= B_i(a) + \sum_{j=1}^{N} p_{ij} \int_{\tau=0}^{\infty} \int_{x=0}^{t} e^{-\lambda x} b_{ij}(x,a) f_{H_{ij}}(\tau,a) dx d\tau \tag{46}$$

where $B_i(a)$ is the immediate cost and is determined by

$$B_i(a) = \sum_{j=1}^{N} p_{ij}(a) B_{ij}(a) \qquad i = 1, \dots, N \qquad B_{ij}(a) < \infty \tag{47}$$

Mathematically, the second term in Eq. (46) represents the sum of the costs that accumulates at the rate per unit time until the transition to state j occurs, $b_{ij}(x,a) < \infty$. For the SMDP model, it is assumed that only 'immediate' costs are incurred when there is a transition from one condition state to another. It can therefore be assumed that the second term in Eq. (46) equals zero. Expressing Eq. (45) in a matrix form, we have [19]:

$$V(a,\lambda) = \left[v_1(a,\lambda) \ \ v(a,\lambda) \ \ \dots \ \ v(a,\lambda) \right]^T \tag{48}$$

$$R(a,\lambda) = V(a,\lambda) = \left[r_1(a,\lambda) \ \ r_2(a,\lambda) \ \ \dots \ \ r_N(a,\lambda) \right]^T \tag{49}$$

$$Q(a,\lambda) = \begin{bmatrix} q_{1,1}(a,\lambda) & q_{1,2}(a,\lambda) & \dots & q_{1,N}(a,\lambda) \\ q_{2,1}(a,\lambda) & q_{2,2}(a,\lambda) & \dots & q_{2,N}(a,\lambda) \\ \dots & \dots & \dots & \dots \\ q_{N,1}(a,\lambda) & q_{N,2}(a,\lambda) & \dots & q_{N,N}(a,\lambda) \end{bmatrix}^T \tag{50}$$

Based on Eq. (45), considering steady state conditions,

$$V(a,\lambda) = R(a,\lambda) + Q(a,\lambda)V(a,\lambda) \tag{51}$$

then,

$$V(a,\lambda) = [I - Q(a,\lambda)]^{-1} R(a,\lambda) \tag{52}$$

As a result, there is a $Q(a,\lambda)$ that gives the minimum long-term cost, which is based on an optimum policy. The minimum long-term cost is defined by Ibe [19] as:

$$v_i^* = min \left\{ r_i(a,\lambda) + \sum_{j=1}^{N} q_{ij}(a,\lambda) v_j^*(a,\lambda) \right\} \qquad i = 1, \dots, N \tag{53}$$

5.5 Do-nothing action (action '*d*')

As mentioned before, for 'do-nothing' actions (a_d) it can be assumed that the sojourn time in a condition state before transition, follows a Weibull distribution, except when sojourning in the terminal state. Using Eq. (39), the numerical solution for the Laplace of the Weibull distribution for each scenario, except failure, can be determined. Let:

$$M_{ij}^{H}(a_d, \lambda) = L\left\{f\left(\infty, a_d | \alpha_{ij}, \beta_{ij}\right)\right\} \quad i,j = 1, \ldots, NS \tag{54}$$

where NS represents the number of states, not including the terminal state. The terminal state can be referred to an absorption condition state or the point at which the bridge element has come to the end of its useful life. In other words, when considering the 'do-nothing' action on a bridge element, once the condition state enters the terminal state, it will remain in that state unless rehabilitated. It follows that,

$$M_{ij}^{H}(a_d, \lambda) = e^{-s} = e^{-\lambda} \quad s = \lambda; \quad i = j = F \tag{55}$$

where F represents the terminal state. From Eqs. (44), (50) and (54),

$$Q(a_d, \lambda) = \begin{bmatrix} p_{1,1} \cdot L\{f(\infty, a_d | \alpha_{1,1}, \beta_{1,1})\} & \cdots & \cdots & p_{1,F} \cdot L\{f(\infty, a_d | \alpha_{1,F}, \beta_{1,F})\} \\ \vdots & \cdots & \cdots & \vdots \\ p_{F,1} \cdot L\{f(\infty, a_d | \alpha_{F,1}, \beta_{F,1})\} & \cdots & \cdots & p_{F,F} \cdot L\{f(\infty, a_d | \alpha_{F,F}, \beta_{F,F})\} \\ 0 & 0 & 0 & e^{-\lambda} \end{bmatrix} \tag{56}$$

where a_d represents 'do-nothing' action.

It can be assumed that the deterioration of bridge elements is a one-step process, and the bridge element can only drop by one condition state at a given point in time. In other words, a bridge element in a particular condition state will eventually transition to the next lower condition state when there are no other actions on that element other than 'do-nothing' actions. It can also assumed that for 'do-nothing' actions, once the condition of the bridge element exits a condition state, that condition state will not be visited again. The transition probability for the embedded Markov chain of the semi-Markov process, p_{ij}, for 'do-nothing' action is therefore assumed to be 1 between the current and succeeding condition states, and for the terminal state. Therefore, for the bridge element that has five (5) condition states plus the terminal state,

$$P_{i,j}(a_d) = \begin{bmatrix} 0 & 1 & 0 & 0 & 0 & 0 \\ 0 & 0 & 1 & 0 & 0 & 0 \\ 0 & 0 & 0 & 1 & 0 & 0 \\ 0 & 0 & 0 & 0 & 1 & 0 \\ 0 & 0 & 0 & 0 & 0 & 1 \\ 0 & 0 & 0 & 0 & 0 & 1 \end{bmatrix} \tag{57}$$

and so Eq. (56) can be simplified to:

$$Q(a_d, \lambda) = \begin{bmatrix} 0 & L\{f(\infty, a_d|\alpha_{1,2}, \beta_{1,2})\} & 0 & 0 & 0 & 0 \\ 0 & 0 & L\{f(\infty, a_d|\alpha_{2,3}, \beta_{2,3})\} & 0 & 0 & 0 \\ 0 & 0 & 0 & \ddots & 0 & 0 \\ 0 & 0 & 0 & 0 & \ddots & 0 \\ 0 & 0 & 0 & 0 & 0 & L\{f(\infty, a_d|\alpha_{5,F}, \beta_{5,F})\} \\ 0 & 0 & 0 & 0 & 0 & e^{-\lambda} \end{bmatrix} \tag{58}$$

5.6 Maintenance action (action '*m*')

For the maintenance action (a_m) the sojourn time distribution can also be Weibull. Each time a maintenance action is done it may affect the condition of each bridge element in the network differently. In other words, the same maintenance action may result in the condition state of a bridge element to either increase, stay the same or decrease at a slower rate than it would have performed if the action was not done. To estimate the transition probability of the embedded Markov chain due to the maintenance action, one can observe the changes that take place in a sample of bridge elements to which the maintenance action was done. To achieve this, the method of least squares using matrices computation can be used to determine the expected transition probability matrix between pairs of observations [32]. Consider Eq. (59):

$$\begin{bmatrix} x_1 & \dots & x_F \\ \vdots & \dots & \vdots \\ x_1^* & \dots & x_F^* \end{bmatrix} \cdot \begin{bmatrix} p_{1,1}(a_m) & \dots & p_{1,F}(a_m) \\ \vdots & \dots & \vdots \\ p_{F,1}(a_m) & \dots & p_{F,F}(a_m) \end{bmatrix} = \begin{bmatrix} y_1 & \dots & y_F \\ \vdots & \dots & \vdots \\ y_1^* & \dots & y_F^* \end{bmatrix} \tag{59}$$

in which each row of x_i's are the proportions of bridge element (in each state) in condition state i before the maintenance action, a_m; each row of y_i's are the proportions of the same bridge element in condition state i after the maintenance action, a_m, for $i = 1, 2, \dots$ NS, F. Each row of the respective matrices with the x_i's and y_i's values represents a different inspection record of a particular bridge element. If the x_i's and y_i's are known for a sample of the same type of bridge element, then the transition probability matrix representing action a_m can be computed using the method of least squares by matrices computations, using Eq. (60). The closest pair of inspection dates before and after a maintenance action can be used to capture the pairs of records having the respective condition states.

$$P_{i,j}(a_m) = \left[X^T X\right]^{-1} \left[X^T Y\right] \tag{60}$$

The resulting transition matrix does not take into consideration the uncertainties as it relates to the time it takes for the maintenance works to be undertaken, which can be described by the distribution of the sojourn time in each condition state. The sojourn time distribution for the maintenance action can also be assumed to follow a Weibull distribution. Using Eq. (39), the numerical solution for the Laplace of the Weibull distribution of the sojourn times irrespective of the current and subsequent condition states can be determined for NS states, excluding the terminal state. Since the solutions for NS states are numerical and the scale (α) and shape (β) estimates are not state specific, then the Laplace of the Weibull distribution for each i,j is represented by:

$$M_{ij}^{H}(a_m, \lambda) = L\{f(\infty, a_m|\alpha, \beta)\} \quad for \ condition \ states \ 1, \ldots, NS \tag{61}$$

where NS represents the number of states, except the terminal state. Eq. (55) can be used to determine $q_{F,F}(a_m, \lambda)$. For a five (5) condition state (plus terminal state) element,

$$Q(a_m, \lambda) = \begin{bmatrix} p_{1,1} \cdot L\{f(\infty, a_m|\alpha, \beta)\} & \cdots & \cdots & \cdots & \cdots & p_{1,F} \cdot L\{f(\infty, a_m|\alpha, \beta)\} \\ p_{2,1} \cdot L\{f(\infty, a_m|\alpha, \beta)\} & \cdots & \cdots & \cdots & \cdots & p_{2,F} \cdot L\{f(\infty, a_m|\alpha, \beta)\} \\ p_{3,1} \cdot L\{f(\infty, a_m|\alpha, \beta)\} & \cdots & \cdots & \cdots & \cdots & p_{3,F} \cdot L\{f(\infty, a_m|\alpha, \beta)\} \\ p_{4,1} \cdot L\{f(\infty, a_m|\alpha, \beta)\} & \cdots & \cdots & \cdots & \cdots & p_{4,F} \cdot L\{f(\infty, a_m|\alpha, \beta)\} \\ p_{5,1} \cdot L\{f(\infty, a_m|\alpha, \beta)\} & \cdots & \cdots & \cdots & \cdots & p_{5,F} \cdot L\{f(\infty, a_m|\alpha, \beta)\} \\ 0 & 0 & 0 & 0 & 0 & e^{-\lambda} \end{bmatrix} \tag{62}$$

5.7 Rehabilitation action (action '*r*')

The duration of particular rehabilitation actions can be determined if sufficient data is available on these rehabilitation projects. If this information is not available, the duration between the two (2) closest inspections prior to the start and subsequent to the completion of the rehabilitation action can be used as an estimate of the duration of the rehabilitation action. The latter was done in this case, assuming two (2) years for the sojourn time. The transition probability of the embedded Markov chain for the rehabilitation action can be assumed to be fixed in which the condition state of the bridge element reverts to 1. As such, the transition probability for the embedded Markov chain of the semi-Markov process, p_{ij}, for rehabilitation action for a five (5) condition state element (plus terminal state) can be represented as:

$$P_{i,j}(a_r) = \begin{bmatrix} 1 & 0 & 0 & 0 & 0 & 0 \\ 1 & 0 & 0 & 0 & 0 & 0 \\ 1 & 0 & 0 & 0 & 0 & 0 \\ 1 & 0 & 0 & 0 & 0 & 0 \\ 1 & 0 & 0 & 0 & 0 & 0 \\ 1 & 0 & 0 & 0 & 0 & 0 \end{bmatrix} \tag{63}$$

Based on Eqs. (40), (44) and (50),

$$Q(a_r, \lambda) = \begin{bmatrix} e^{-2\lambda} & 0 & 0 & 0 \\ e^{-2\lambda} & 0 & 0 & 0 \\ e^{-2\lambda} & 0 & 0 & 0 \\ e^{-2\lambda} & 0 & 0 & 0 \\ e^{-2\lambda} & 0 & 0 & 0 \\ e^{-2\lambda} & 0 & 0 & 0 \end{bmatrix} \tag{64}$$

6. Conclusions

This chapter outlined feasible approaches that demonstrate how the condition of transportation infrastructure can be used to develop Markov and semi-Markov deterioration models that are able to model network level performance when no actions are done to the infrastructure. The use of semi-Markov processes to model deterioration relaxes the assumption of the distribution of the sojourn time in condition states for 'do-nothing' actions and is therefore less restrictive than Markov chain deterioration models. A preservation model for transportation infrastructure using Semi-Markov Decision Processes was presented. The objective of the preservation model is to determine the minimum long-term costs for the preservation of a transportation infrastructure within a group or network of similar infrastructure. The main 'inputs' for the preservation model are: (a) the scale and shape parameter estimates of the Weibull distribution used to describe 'do-nothing' actions, and (b) the transition probabilities of the embedded Markov chain and the scale and shape parameter estimates of the Weibull distribution for maintenance actions. If there is sufficient data available, the use of the Semi-Markov Decision Process is a useful tool to use for modeling Transportation Infrastructure preservation.

Acknowledgements

I would like to acknowledge and express my gratitude to my wife, Tamika, and my children, Aiden and Grace, for their support and the encouragement during the process of writing this chapter. I must say thanks to Almighty God for His continuous grace and guidance!

Nomenclature

CRK	Crack Index
NS	number of states

Abbreviations

PMS	Pavement Management System
ADOT	Arizona Department of Transportation
NOS	Network Optimization System
MLE	Maximum Likelihood Estimation
iid	identical and independent distributed
pdf	probability density function
pmf	probability mass function
MDP	Markov Decision Process
SMDP	Semi-Markov Decision Process
BMS	Bridge Management System
MR&R	maintenance, repair and rehabilitation

Author details

Omar Thomas[1*] and John Sobanjo[2]

1 The University of the West Indies, Kingston, Jamaica

2 FAMU-FSU College of Engineering, Tallahassee, Florida, USA

*Address all correspondence to: omar.thomas@uwimona.edu.jm

References

[1] Wang CP. Pavement network optimization and analysis [PhD dissertation]. Tempe, AZ, USA: Arizona State University; 1992

[2] Wang KCP, Zaniewski J, Way G. Probabilistic behavior of pavements. Journal of Transportation Engineering, American Society of Civil Engineers (ASCE). 1994;**120**(3):358-375

[3] Nasseri S, Gunaratne M, Yang J, Nazef A. Application of improved crack prediction methodology in florida's highway network. Transportation Research Record: Journal of the Transportation Research Board. 2009; **2093**:67-75

[4] Golabi K, Thompson PD, Hyman WA. Pontis Version 2.0 Technical Manual, A Network Optimization System for Bridge Improvements and Maintenance. Washington DC: Federal Highway Administration; December 1993

[5] Micevski T, Kuczera G, Coombes P. Markov model for storm water pipe deterioration. Journal of Infrastructure Systems, American Society of Civil Engineers (ASCE). 2002;**8**(2):49-56

[6] Baik H-S, Seok HJ, Abraham DM. Estimating transition probabilities in markov chain-based deterioration models for management of wastewater systems. Journal of Water Resources Planning and Management, American Society of Civil Engineers (ASCE). 2006; **132**(1):15-24

[7] Yang J, Gunaratne M, Jian John L, Dietrich B. Use of recurrent markov chains for modeling the crack performance of flexible pavements. Journal of Transportation Engineering, American Society of Civil Engineers (ASCE). 2005;**131**(11):861-872

[8] Yang J, Lu JJ, Gunaratne M, Dietrich B. Modeling crack deterioration of flexible pavements: Comparison of recurrent markov chains and artificial neural networks. Transportation Research Record: Journal of the Transportation Research Board. 1974;**18–25**:2009

[9] Ng S-K, Moses F. Bridge deterioration modeling using semi-markov theory. A. A. Balkema Uitgevers B.V. Structural Safety and Reliability. 1998;**1**:113-120

[10] Sobanjo JO. State transition probabilities in bridge deterioration based on weibull sojourn times. Structure and Infrastructure Engineering: Maintenance, Management, Life-Cycle Design and Performance. 2011;**7**(10):747-764

[11] Black M, Brint AT, Brailsford JR. Comparing probabilistic methods for the asset management of distributed items. Journal of Infrastructure Systems, American Society of Civil Engineers (ASCE). 2005;**11**(2):102-109

[12] Black M, Brint AT, Brailsford JR. A semi-markov approach for modelling asset deterioration. Journal of the Operational Research Society. 2005;**56**: 1241-1249

[13] Thomas O. Stochatic preservation model for transportation infrastructure [PhD dissertation]. Tallahassee, FL, USA: Florida State University; 2011

[14] Ross SM. Stochastic Processes. 2nd ed. USA: John Wiley and Sons, Inc.; 1996

[15] Casella G, Berger RL. Statistical Inference. 2nd ed. Duxbury, Pacific Grove, CA, USA: Cenage Learning; 2001

[16] Billington R, Allan RN. Reliability Evaluation of Engineering Systems: Concepts and Techniques. London: Pitman Books Limited; 1983

[17] Tobias PA, Trindade DC. Applied Reliability. 2nd ed. Florida: Chapman and Hall/CRC Press; 1995

[18] Birolini A. Reliability Engineering, Theory and Practice. 5th ed. Berlin, Heidelberg: Springer, Verlag; 2007

[19] Ibe OC. Markov Processes for Stochastic Modeling. Massachusetts: Elsevier Academic Press; 2009

[20] Howard RA. Dynamic probabilistic systems. In: Volume II: Semi-Markov and Decision Processes. Canada: John Wiley and Sons Inc.; 1971

[21] Cleves MA, Gould WW, Guitierrez RG. An Introduction to Survival Analysis Using Stata. Revised ed. 4905 Lakeway Drive, College Station, Texas 77845: Stata Press; 2001

[22] Castillo E, Hadi AS, Balakrishnan N, Sarabia JM. Extreme Value and Related Models with Applications in Engineering and Science. New Jersey: John Wiley and Sons, Inc.; 2005

[23] Lee ET. Statistical Methods for Survival Data Analysis. 2nd ed. USA: John Wiley and Sons, Inc.; 1992

[24] Durango PL, Madanat SM. Optimal maintenance and repair policies in infrastructure management under uncertain facility deterioration rates: And adaptive approach. Transportation Research Part A. 2002;**36**:763-778

[25] Guillaumot VM, Durango-Cohen PL, Madanat SM. Adaptive optimization of infrastructure maintenance and inspection decisions under performance model uncertainty. Journal of Infrastructure Systems, American Society of Civil Engineers (ASCE). 2003; **9**(4):133-139

[26] Gonzalez J, Romera JCR, Perez JM. Optimal railway infrastructure maintenance and repair policies to manage under uncertainty with adaptive control. UC3M Working Papers. Statistics and Econometrics. 2006;**5**:1-15

[27] Madanat SM, Park S, Kuhn K. Adaptive optimization of infrastructure maintenance and inspection decisions under performance model uncertainty. Journal of Infrastructure Systems, American Society of Civil Engineers (ASCE). 2006;**12**(3):192-198

[28] Thompson PD, Harrison FD. Pontis Version 2.0 User's Manual, A Network Optimization System for Bridge Improvements and Maintenance. Washington DC: Federal Highway Administration; December 1993

[29] Evans M, Hastings N, Peacock B. Statistical Distributions. 3rd ed. New York: John Wiley and Sons; 2000

[30] Puterman ML. Markov Decision Processes: Discrete Stochastic Dynamic Programming. New Jersey: John Wiley and Sons; 2005

[31] Sagias NC, Karagiannidis GK. Gaussian class multivariate weibull distributions: Theory and applications in fading channels. Institute of Electrical and Electronics Engineers. Transactions on Information Theory. 2005;**51**(10): 3608-3619

[32] Thompson PD, Johnson MB. Markovian bridge deterioration: Developing models from historical data. Structure and Infrastructure Engineering. 2005;**1**:85-91

Chapter 5

Some Applications of the Partially Observable Markov Decision Processes

Doncho S. Donchev

Abstract

We give two examples of reduction of a POMDPs model to a MDP model with a complete information. In both MDP models, corresponding Bellman's equations allow to construct optimal policies. The first example is the classical discrete time disorder problem. We include it into the framework of POMDPs in contrast to the standard approach based on optimal stopping rules. In the second model, we consider the steps of a transaction under threat as a trajectory of POMDP. We show how to define the elements of the model in order to achieve an optimal balance between the successful end of the transaction and minimum applications of mitigating actions.

Keywords: transaction models, intrusion detection and information filtering, cost functions, partially observable Markov decision processes, disorder problem, Bellman's equation

1. Introduction

Methods based on the Markov decision process (MDP) have become an essential part of the reinforcement learning in environments in which we can observe directly the evolution of the controlled object. In many situations, however, we suffer from limited capabilities to do that. The partially observable Markov decision processes (POMDPs) extend the MDP framework allowing us for decision making under the conditions of uncertainty. The classical approach to POMDPs suggests that in order to find the optimal policy in that model one should perform its reduction to a MDP model with complete information which consists of the following steps:

1. Find suitable sufficient statistics that play the role of a controlled process in the MDP model with complete information;

2. Define all elements of the MDP model;

3. Solve Bellman's equation for the model from step 2.

Unfortunately, there is a lack of examples for which this procedure can be completed successfully. In this chapter, we present two POMDPs models by focusing on the algorithm described above and show how the optimal policies for these POMDPs can be represented.

Firstly, we revisit the discrete time disorder problem which was stated and investigated in details by Shiryaev [1]. The formal setting of this problem is the following. On some measurable space $(\Omega, \mathcal{F})$ are given random variables $\theta, \xi_1, \xi_2, \ldots$ and a probability measure P^π such that

$$P^\pi(\theta = 0) = \pi, P^\pi(\theta = n) = (1 - \pi)p(1 - p)^{n-1}, \quad n \geq 1,$$

where $0 < p < 1, 0 \leq \pi \leq 1$ are some constants, and the probability of any event $A = \{\omega : \ \xi_1 \leq x_1, \ldots, \xi_n \leq x_n\}$ is equal to

$$P^\pi(A) =$$

$$= \pi P^1(A) + (1 - \pi) \sum_{i=0}^{n-1} p(1 - p)^i P^0\{\xi_1 \leq x_1, \ldots, \xi_i \leq x_i\}$$

$$\times \quad P^1\{\xi_{i+1} \leq x_{i+1}, \ldots, \xi_n \leq x_n\} + (1 - \pi)(1 - p)P^0(A).$$

In the last formula both measures P^0 and P^1 define c.d.fs. $F_i(x) = P^i(\xi_1 \leq x), \ i = 0, 1$ with densities $p_i(x), \ i = 0, 1$. W.r.t. $P^i, \ i = 0, 1$ all random variables $\xi_1, \xi_2, \ldots$ are i.i.d.

Let $\mathcal{F}^\xi = \{\mathcal{F}_n^\xi\}, \mathcal{F}_0^\xi = \{\phi, \Omega\}, \mathcal{F}_n^\xi = \sigma(\xi_1, \xi_2, \ldots, \xi_n), \ n = 1, 2, \ldots$ be a filtration adapted to the observations $\xi_1, \xi_2, \ldots$. The problem is to find a $\mathcal{F}^\xi$-stopping time τ which minimizes the risk

$$\rho^\pi(\tau) = P^\pi(\tau < \theta) + cE^\pi(\tau - \theta)^+,$$

where c is a positive constant. A solution to that problem in terms of optimal stopping rules and martingale techniques can be found in [1].

In Section 3, we present a different approach to this problem including it into the framework of POMDPs. We show that steps 1 and 2 of the reduction can be performed successfully. Moreover, we define asymptotic solutions to Bellman's equation and describe a procedure which allows to find them for some densities $p_0(x)$ and $p_1(x)$.

Section 4 is devoted to applications of POMDPs in the cybersecurity. The Contemporary Intrusion Detection Systems (IDS) are widely used in network management and cybersecurity frameworks for detecting and classifying potential security threats of unauthorized intrusions. The unauthorized intrusions during transaction processing are particularly dangerous because they can lead to significant financial losses. There are a number of security measures which can be used to counteract, but the success of their use depends on the precision of the detection and the time. In both network-based IDS, which use signature information [2], and host-based IDS, which use behavioral information [3] the data analysis utilizes a variety of methods for Machine Learning (ML).

In our recent papers, see [4, 5], we developed an approach based on the POMDPs which allows to perform a quantitative assessment of the impact of the parameters of the model on different important characteristics of the transaction. A deficiency of the model presented in [4, 5] is the assumption for a constant intrusion rate. Here, we relax it. Moreover, we reveal the crucial role of the costs of counteractions in building the model. On the one hand, they should favor the successful end of the transaction, but, on the other hand, prevent the unnecessary use of counteractions when there is no an attack. We show that the costs can be found as optimal solutions of some linear program. We complete this section by solving Bellman's equation and describing the optimal policy in the model.

In order to make this chapter selfcontained we make a general description of POMDPs models in Section 2.

2. A general description of models with incomplete information

We remind the general framework of the model with incomplete information for controlled Partially Observable Markov Decision Processes (POMDP). For more details we refer the reader to [6, 7]. The model consists of the following elements:

1. The state space X. A non-empty Borel space.

2. An action space A. A non-empty Borel space.

3. An observation space Z. A non-empty Borel space.

4. A transition kernel $t(dx_{n+1}|x_n, a_n)$. A probability measure on X depending on both the last state x_n and the last action a_n.

5. An observation kernel $s(dz_{n+1}|x_{n+1}, a_n)$. A probability measure on Z depending on both the current state x_{n+1} and the last action a_n.

6. An initial observation $s_0(dz_0|x_0)$.

7. A running reward function $g(x_n, a_n)$.

8. A horizon N. A positive integer or ∞. If $N < \infty$ we introduce also a final reward $R(x_N)$.

We assume that all transition kernels and rewards are measurable.

The evolution of the POMDP is the following. At time n, $n = 0,1,2, \ldots$ it jumps from state x_n to state x_{n+1} according to the law $t(dx_{n+1}|x_n, a_n)$. We do not observe both x_n and x_{n+1}. Instead, we observe z_{n+1}, whose distribution is $s(dz_{n+1}|x_{n+1}, a_n)$, and add this observation to the information available till time $n+1$: $i_{n+1} = z_0a_0 \ldots z_na_nz_{n+1}$. The initial information i_0 coincides with the initial observation z_0 which depends on the initial state x_0. We assume that the initial distribution p of x_0 is also known.

Definition 1 A policy is a sequence $\pi = (\mu_0, \mu_1, \ldots, \mu_{N-1})$ of measures $\mu_n(da_n|p, i_n)$ such that $\mu_n(A_n|p, i_n) = 1$, where A_n is the set of all admissible actions corresponding to the initial distribution p and the information i_n.

Besides the information i_n, we consider also the set of all histories H_n till time n. Each history $h_n \in H_n$ includes not only i_n but also all states visited by the process till time n : $h_n = x_0z_0a_0x_1z_1a_1 \ldots x_nz_na_n$. According to Ionesku-Tulchea's theorem to each initial distribution p and policy π on the set H_n there corresponds a measure $P_n^\pi(p)$ given by the formula:

$$P_n^{\pi}(p)(X_0Z_0A_0 \dots X_nZ_nA_n) =$$
$$= \int_{X_0}\int_{Z_0}\int_{A_0} \cdots \int_{X_n}\int_{Z_n} \mu_n(A_a|p,i_n) \times$$
$$\times s(dz_n|x_n,a_{n-1})t(dx_n|x_{n-1},a_{n-1} \dots \mu_0(da_0|p,z_0)s_0(dz_0|x_0)p(dx_0).$$

Here $X_0Z_0A_0 \dots X_nZ_nA_n$ is a cylindrical subset of H_n. The problem is to find a policy π which maximizes the total reward

$$J_N^{\pi}(p) = E_N^{\pi}(p)\left(\sum_{n=0}^{N-1} g(x_n,a_n) + R(x_N)\right), \tag{1}$$

where $E_N^{\pi}(p)$ is an expectation w.r.t. the measure $P_N^{\pi}(p)$, and the term $R(x_N)$ is missing if $N = \infty$.

If $N = \infty$, then in order to ensure convergence in (1) often introduce a discounting. However, the infinite horizon model which we study here is such that in (1) with probability 1 there are a finite number nonzero terms. So, we do not need to take care for the convergence.

The general approach to the optimal control of POMDP consists in a reduction of problem (1) to a MDP with a complete information. We pay no attention to that reduction in a full generality, referring the reader to [6, 7]. However, in the next section we consider in details all steps of the reduction.

3. The disorder problem

In this section we develop an approach to the discrete time disorder problem based on POMDPs. We define the elements of the general model described in Section 2 as follows:

1. The state space $X = \{0,1,2\}$ consists of three states $0, 1$ and 2.

2. The action space $A = \{wait, stop\}$ consists of the two possible actions we can apply.

3. The observation space Z coincides with the support of both densities $p_0(z)$ and $p_1(z)$. We assume that the supports of $p_0(z)$ and $p_1(z)$ coincide for, otherwise, in some cases we will observe directly the current state of the controlled process.

4. We define the transition kernel t by:

$$t(1|0,wait) = p, \;\; t(0|0,wait) = 1-p, \;\; t(1|1,wait) = 1,$$
$$t(2|0,stop) = t(2|1,stop) = t(2|2) = 1.$$

5. The observation kernel is

$$s(dz|0,wait) = p_0(z)dz, s(dz|1,wait) = p_1(z)dz.$$

6. The running reward is:

$$g(0, wait) = 1, \;\; g(1, wait) = -c < 0, g(0, stop) = g(1, stop) = 0, \;\; g(2) = 0.$$

7. The initial distribution on X is

$$P(x_0 = 0) = 1 - \pi, \;\; P(x_0 = 1) = \pi, \;\; P(x_0 = 2) = 0.$$

8. The horizon $N = \infty$.

Let us clarify how this model relates to the disorder problem as stated above. The moment of disorder θ which is $Ge(p)$-distributed is in fact the time of the first visit of the controlled process at state 1. Our goal is to fix θ as precisely as possible making use of the action *stop*. If we are in state 0 then the disorder still not occur, and the right decision is to wait. This yields an income equal to 1, and our total reward increases by the same amount. However, if we miss θ (and therefore being in state 1) then applying the action *wait* is certainly a wrong decision which results in paying a positive penalty $c > 0$. The only way to avoid such a situation is to stop the process, and to go to the absorbing state 2. Since we apply the action *stop* once, say at time τ, we can identify any policy with that time, and consider the following problem: Maximize the total expected reward

$$J^{\tau}(\pi) = E^{\pi} \sum_{n=0}^{\tau-1} g(x_n, wait), \tag{2}$$

where E^{π} is an expectation corresponding to the initial distribution given by Element 7 of the model.

Remark. In contrast to the optimal stopping approach where we receive our income when applying the action *stop*, formula (2) shows that now the situation is completely different—the whole reward is a result of waiting actions whereas the action *stop* yields no income.

3.1 Filtration equations. Sufficient statistics. Bellman's equation

The total probability formula implies the following representation of $J^{\tau}(\pi)$:

$$J^{\tau}(\pi) = (1 - \pi) J^{\tau}(0) + \pi J^{\tau}(1). \tag{3}$$

Evidently, $J^{\tau}(1) = 0$ because, being aware that the process starts at state 1, the optimal policy recommends an immediate stop. In order to find $J^{\tau}(0)$ we follow a standard procedure of reducing the POMDP model to a model with complete information.

The first step of this procedure consists in solving the filtration equations. Denote by $f_n(i), \;\; i = 0, 1, n = 1, 2, \ldots$ (resp. $r_n(i), \;\; i = 0, 1, \;\; n = 1, 2, \ldots$) the prior (resp. posterior) probabilities of states $i, \;\; i = 0, 1$ at time n. Namely, $f_n(i)$ is the probability of being in state i before observing ξ_n whereas $r_n(i)$ is the same probability after the observation has occur. Let $\xi_1 = z$. Since at the beginning we certainly know that the process starts at state 0, and applying the Bayes rule we get

$$f_1(0) = 1 - p, \; f_1(1) = p, \tag{4}$$

$$r_1(0) = \frac{f_1(0)p_0(z)}{f_1(0)p_0(z) + f_1(1)p_1(z)}, r_1(1) = \frac{f_1(1)p_1(z)}{f_1(0)p_0(z) + f_1(1)p_1(z)}. \tag{5}$$

Similarly, under the assumption that $\xi_n = z$, for any $n > 1$ one holds

$$f_n(0) = (1-p)r_{n-1}(0), f_n(1) = pr_{n-1}(0) + r_{n-1}(1), \tag{6}$$

$$r_n(0) = \frac{f_n(0)p_0(z)}{f_n(0)p_0(z) + f_n(1)p_1(z)}, r_n(1) = \frac{f_n(1)p_1(z)}{f_n(0)p_0(z) + f_n(1)p_1(z)}. \tag{7}$$

The general theory recommends using suitable posterior probabilities, say $r_n(1), \; n = 1, 2, \ldots$, as sufficient statistics. However, in our case it is more convenient to introduce other sufficient statistics. Indeed, formulas (6) and (7) imply that

$$\frac{f_n(1)}{f_n(0)} = \frac{pr_{n-1}(0) + r_{n-1}(1)}{(1-p)r_{n-1}(0)} = \frac{p}{1-p} + \frac{1}{1-p}\frac{r_{n-1}(1)}{r_{n-1}(0)}. \tag{8}$$

$$\frac{r_n(1)}{r_n(0)} = \frac{p_1(z)}{p_0(z)}\frac{f_n(1)}{f_n(0)} = \frac{p_1(z)}{p_0(z)}\left(\frac{p}{1-p} + \frac{1}{1-p}\frac{r_{n-1}(1)}{r_{n-1}(0)}\right). \tag{9}$$

Let us set

$$X_n = \frac{r_n(1)}{pr_n(0)} + 1, n = 1, 2, \ldots \tag{10}$$

Then, (17) reduces to

$$X_n = X_{n-1}/a(z) + 1, a(z) = \frac{(1-p)p_0(z)}{p_1(z)}. \tag{11}$$

The linear relation between X_n and X_{n-1} in (11) makes the process (X_n) very convenient to work with, and suitable to be the sufficient statistics for our problem.

In order to complete the reduction to a fully observable MDP it remains to express the reward g_n in terms of X_n, and to find the conditional distribution of ξ_n provided that $X_{n-1} = y$. Making use of (10), and the fact that $r_n(0) + r_n(1) = 1$, we get

$$r_n(0) = \frac{1}{1 + p(X_n - 1)}, r_n(1) = \frac{p(X_n - 1)}{1 + p(X_n - 1)}. \tag{12}$$

On the other hand, the expected reward of the waiting action at step n is equal to $r_n(0) - cr_n(1)$. Thus, in view of (12), the reward in the reduced MDP model should be

$$g(X_n) = 1 - (c+1)\frac{p(X_n - 1)}{1 + p(X_n - 1)}. \tag{13}$$

Comparing (8) and (10) we get

$$\frac{f_n(1)}{f_n(0)} = \frac{p}{1-p} X_{n-1}. \tag{14}$$

Making use of the first equalities in (6), (12), as well as (14), we obtain

$$\begin{aligned} P(\xi_n \in dz | X_{n-1} = y) &= dz(f_n(1)p_1(z) + f_n(0)p_0(z)) \\ &= dzf_n(0)\left(\frac{f_n(1)}{f_n(0)}p_1(z) + p_0(z)\right) \\ &= dzf_n(0)\left(\frac{py}{1-p}p_1(z) + p_0(z)\right) \\ = dz\left(\frac{py}{1+p(y-1)}p_1(z) + \frac{1-p}{1+p(y-1)}p_0(z)\right). \end{aligned} \tag{15}$$

Thus, in the reduced fully observable MDP model the controlled process is the Markov chain $X_n = X_{n-1}/a(\xi_n) + 1$. Its transitions are according to formula (15), and each waiting action yields an income (3). The action *stop* yields no income and terminates the process. We want to maximize the total reward

$$R^\tau(y) = E\left(\sum_{n=1}^{\tau-1} g(X_n) | X_1 = y\right), \sum_1^0 = 0,$$

where τ is the time when we apply the action *stop*. Let us set

$$V(y) = \sup_\tau R^\tau(y).$$

The function $V(y)$ satisfies Bellman's equation

$$\begin{aligned} V(y) &= \left(1 - (c+1)\frac{p(y-1)}{1+p(y-1)}\right. \\ &+ \left.\int_Z V(y/a(z)+1)\left(\frac{pyp_1(z)}{1+p(y-1)} + \frac{(1-p)p_0(z)}{1+p(y-1)}\right)dz\right)^+. \end{aligned} \tag{16}$$

It is strictly positive on some interval $(1, y_0)$, $y_0 = y_0(c,p)$, and is equal to zero if $y \geq y_0$. If the threshold y_0 is known, then the optimal stopping rule is

$$\tau = \min(n \geq 1 : X_n > y_0).$$

3.2 Asymptotic solutions to Bellman's equation

Multiplying (16) by $1 + p(y-1)$ we get

$$(1+p(y-1))V(y) = \left(1 - cp(y-1) + \int_Z V\left(\frac{y}{a(z)} + 1\right)(pyp_1(z) + (1-p)p_0(z))dz\right)^+. \tag{17}$$

Making use of representation $py = (1+py/a(z)-1)a(z)$, and taking into consideration the second equality in (11) we see that the integral in the right-hand side of (17) is equal to

$$\int_Z V\left(\frac{y}{a(z)}+1\right)(1+py/a(z))a(z)p_1(z)dz.$$

The substitution $a(z)=t$ reduces the last integral to

$$\int_{a(Z)} V\left(\frac{y}{t}+1\right)\left(1+p\frac{y}{t}\right)t^2(p_1{\circ}a)(t)\frac{dt}{t}, \tag{18}$$

where $(p_1{\circ}a)(t)=p_1(a^{-1}(t))/a'(a^{-1}(t))$ is the transformation of the density p_1 with the function $a(t)$.

Let us introduce a new unknown function $U(y)=(1+p(y-1))V(y)$. Then, in terms of $U(y)$, and in view of (18), Eq. (17) reduces to

$$U(y)=\left(1-cp(y-1)+\int_{a(Z)} U\left(\frac{y}{t}+1\right)t^2(p_1{\circ}a)(t)\frac{dt}{t}\right)^+. \tag{19}$$

Let $c=0$. Then, the optimal policy does not recommend using the action *stop* at all, since then both states 1 and 2 become indistinguishable and we can only lose if the process is in state 0. Therefore, in this case the function U satisfies the equation

$$U(y)=1+\int_{a(Z)} U\left(\frac{y}{t}+1\right)t^2(p_1{\circ}a)(t)\frac{dt}{t}.$$

This observation motivates us to introduce asymptotic solutions to Eq. (16).

Definition 2 The function $V(y)=(U(y)/(1+p(y-1)))^+,\ y\geq 1$ is an asymptotic solution to the eq. (16) as $c\to 0$, if the function $U(y)$ satisfies the equation

$$U(y)=1-cp(y-1)+\int_{a(Z)} U\left(\frac{y}{t}+1\right)t^2(p_1{\circ}a)(t)\frac{dt}{t}. \tag{20}$$

The asymptotic optimal threshold y_0 is the root of the equation $U(y)=0$.

We are going to apply Mellin's transform to Eq. (20). In order to be consistent with that transform, we assume that the set $a(Z)$ (and therefore Z) does not intersect with the negative part of the real line.

Let us recall that the Mellin transform $M[f](s)$ of function $f(x), x>0,$ is defined by the formula

$$M[f](s)=\int_0^\infty x^{s-1}f(x)dx,\ \ s\in\Omega_f. \tag{21}$$

It is an analytical function inside domain Ω_f which is usually either a half-plane or a stripe of the complex plane parallel to the imaginary axis. The Mellin transform is invertible and the inversion formula is

$$f(x) = \frac{1}{2\pi i}\int_{\gamma-i\infty}^{\gamma+i\infty} M[f](s)x^{-s}ds, \gamma \in \Omega_f. \tag{22}$$

We are looking for solutions to Eq. (20) which belong to the class of Meyer's G-functions. These functions have the property of their Mellin's transform being

$$\Gamma\begin{bmatrix} b_1+s, \cdots, b_m+s, 1-a_1-s, \cdots, 1-a_n-s \\ a_{m+1}+s, \cdots, a_r+s, 1-b_{m+1}-s, \cdots, 1-b_q-s \end{bmatrix}$$

$$:= \frac{\prod_{i=1}^{m}\Gamma(b_i+s)\prod_{i=1}^{n}\Gamma(1-a_i-s)}{\prod_{i=m+1}^{r}\Gamma(a_i+s)\prod_{i=m+1}^{q}\Gamma(1-b_i-s)}, \tag{23}$$

where $\Gamma(\cdot)$ is the Euler gamma function, $0 \leq m \leq q, 0 \leq n \leq r$, and a_i and $b_j, i = 1, \dots, r, j = 1, \dots, q$ are numbers. Formally, Meyer's G-function is given by the integral

$$G_{rq}^{mn}(z|(a_r);(b_q)) = G_{rq}^{mn}(z|a_1, \dots, a_r; b_1, \dots, b_q) =$$

$$\frac{1}{2\pi i}\int_{\gamma-i\infty}^{\gamma+i\infty}\Gamma\begin{bmatrix} b_1+s, \dots, b_m+s, 1-a_1-s, \dots, 1-a_n-s \\ a_{m+1}+s, \dots, a_r+s, 1-b_{m+1}-s, \dots, 1-b_q-s \end{bmatrix} z^{-s}ds, \tag{24}$$

where γ is a number such that all poles $-b_j-k,\ k=0,1,\dots$ (resp. $1-a_i+k,\ k=0,1,\dots$) remain in the left (resp. right) half-plane w.r.t. the line $Re(s)=\gamma$. The properties of Meyer's G-function as well as tables containing representations of many elementary and special functions in terms of Meyer's G-functions can be found in [8], Chapter 8.

The following formulas represent Meyer's G-functions in terms of generalized hypergeometric functions:

$$G_{rq}^{mn}\left(z\middle|\begin{matrix}(a_r)\\(b_q)\end{matrix}\right)$$

$$= \sum_{k=1}^{m}\Gamma\begin{bmatrix} b_1-b_k, \hat{\cdots}, b_m-b_k, 1+b_k-a_1, \dots, 1+b_k-a_n \\ a_{n+1}-b_k, \dots, a_r-b_k, 1+b_k-b_{m+1}, \dots, 1+b_k-b_q \end{bmatrix} z^{b_k}$$

$$\times {}_rF_{q-1}\left(1+b_k-(a_r); 1+b_k-(b_q)'; (-1)^{r-m-n}z\right), \tag{25}$$

$$r \leq q,\ b_j - b_k \neq 0, \pm 1 \pm 2, \dots$$

$$= \sum_{k=1}^{n}\Gamma\begin{bmatrix} a_k-a_1, \hat{\cdots}, a_k-a_n, 1+b_1-a_k, \dots, 1+b_m-a_k \\ a_k-b_{m+1}, \dots, a_k-b_q, 1+a_{n+1}-a_k, \dots, 1+a_r-a_k \end{bmatrix} z^{a_k-1}$$

$$\times {}_qF_{r-1}\left(1+(b_q)-a_k; 1+(a_r)'-a_k; (-1)^{s-m-n}z^{-1}\right) \tag{26}$$

$$r \geq q,\ a_j - a_k \neq 0, \pm 1, \pm 2, \dots$$

Here, $(a_r)' - a_k = a_1 - a_k, \dots, a_{k-1} - a_k, a_{k+1} - a_k, \dots, a_r - a_k$ and $\Gamma[a_1 - a_k, \hat{\cdots}, a_m - a_k] := \Gamma(a_1 - a_k) \dots \Gamma(a_{k-1} - a_k)\Gamma(a_k + 1 - a_k) \dots \Gamma(a_m - a_k)$.

In order to apply the Mellin transform to Eq. (12), we need the following lemma for the Mellin transform of the composition of two functions.

Lemma 1 Let $f(x)$ and $\varphi(x)$ be two functions, $F(s) = M[f](s)$ and assume that the Mellin transform of function $(\varphi(x))^{-u}$ exists for all u from the line $Re\,(u) = \beta$ which belongs to Ω_f. Let us set $G(s,u) = M[(\varphi(x))^{-u}](s)$. Then

$$M[f(\varphi(x))](s) = \frac{1}{2\pi i}\int_{\beta-i\infty}^{\beta+i\infty} G(s,u)F(u)du. \tag{27}$$

Proof. We apply the inversion formula (22) to the right-hand side of (27) taking γ to be such that line $Re\,(s) = \gamma$ belongs to $\Omega_{\varphi^{-u}}$. Changing the order of integration, we get

$$\frac{1}{2\pi i}\int_{\gamma-i\infty}^{\gamma+i\infty} x^{-s}ds\frac{1}{2\pi i}\int_{\beta-i\infty}^{\beta+i\infty} G(s,u)F(u)du$$

$$= \frac{1}{2\pi i}\int_{\beta-i\infty}^{\beta+i\infty} F(u)du\int_{\gamma-i\infty}^{\gamma+i\infty} G(s,u)x^{-s}ds$$

$$= \frac{1}{2\pi i}\int_{\beta-i\infty}^{\beta+i\infty} F(u)(\varphi(x))^{-u}du = f(\varphi(x)),$$

which ends the proof.

Let us set $F(s) = M[U](s),\ \ A(s) = M\left[t^2(p_1\circ a)(t)\right](s)$. Applying Mellin's transform to (20), and making use of the formula for the transform of a convolution, we get

$$F(s) = \frac{cp}{s+1} - \frac{1+cp}{s} + M[W(y+1)](s)A(s). \tag{28}$$

Applying Lemma 1 to both functions $U(y)$ and $y+1$ we obtain

$$M[W(y+1)](s) = \frac{1}{2\pi i}\int_{\beta-i\infty}^{\beta+i\infty} B(s,u-s)F(u)du,$$

where $B(s,u-s)$ is the beta-function, and β is such that the line of integration belongs to Ω_U. Substituting the last formula in (28) we come to the equation

$$F(s) = \frac{cp}{s+1} - \frac{1+cp}{s} + \frac{1}{2\pi i}\int_{\beta-i\infty}^{\beta+i\infty} B(s,u-s)F(u)duA(s). \tag{29}$$

If $U(y)$ is a G-function, then $F(s)$ is of the form (23), and so is the function under the integral in (29). The integral itself can be found by formula (24), and evaluated according to formulas (25) and (26). Thus, we can find $A(s)$, $(p_1\circ a)(t)$, and finally densities $p_0(z)$ and $p_1(z)$ which correspond to the function $(p_1\circ a)(t)$. The function $V(y) = (U(y)/(1+p(y-1)))^+$ will be an asymptotic solution to Eq. (16), corresponding to these densities.

This approach has been applied in [9], where we have found the stationary distribution of the Markov chain $X_n = \xi_n X_{n-1} + 1$ in case of some special i.i.d. random variables $\xi_n, n = 1, 2, \ldots$.

4. The role of the cost functions for preventing intrusions in transactions under threat

Here, we briefly describe the model studied in [4, 5] which we denote by M. It consists of the following elements:

1. **State space** $S = \{safe, danger, deadend\}$—corresponds to the different types of situations from the point of view of the risk they pose:

   a. *safe* Situations along the transactions in absence of any threats;

   b. *danger* Situations in which the system is under the influence of security threats but still able to recover and

   c. *deadend* Situations in which the system experiences severity and crashes completely under the security threats.

2. **Action space** $A = \{noact, respond\}$—corresponds to the two types of actions from risk perspective:

   a. *noact*—no action intervention, the system goes straight to the next situation according to the recommended action and continues the normal track of execution of the current transaction, and

   b. *respond*—counteraction, which brings the system back to a safe situation after malicious action deviating the transaction from its normal course.

3. **Observation space** $Z = \{nothreat, threat, crash\}$—corresponds to the different types of events from risk viewpoint:

   a. *nothreat*—asynchronous event, which is non-threatening and does not require counteraction;

   b. *threat*—detection of malicious intervention which requires counteraction, and

   c. *crash*—losing control of the system without chance for recovery.

4. **Transition kernel** $q(s_{n+1}|s_n, a_{n+1})$—probability of the transition from situation s_n to situation s_{n+1} under control c_{n+1}, calculated as follows:

   - $q(safe|safe) = p$, p is the probability for absence of threats after transition from a safe situation;

   - $q(danger|safe) = 1 - p$, $1 - p$ is the probability for presence of threats after transition to from safe situation;

   - $q(safe|danger, respond) = 1$ because the counteraction in dangerous situation eliminates the threat;

- $q(deadend|danger, noact) = 1$ because the absence of counteraction in dangerous situation leads to inevitable crash of the system, and
- $q(deadend|deadend) = 1$ since there is no way out of the crash.

5. **Occurrence kernel** $t(z_n|s_n)$—probability of the occurrence of event z_n in situation s_n, calculated as follows:
   - $t(nothreat|safe) = p_{11}$—probability of not observing threat occurrence in a safe state (*true negative*);
   - $t(nothreat|danger) = p_{12}$—probability of not observing threat occurence in a dangerous stage (*false negative*);
   - $t(threat|safe) = p_{21}$—probability of observing threat occurrence in a safe state (*false positive*);
   - $t(threat|danger) = p_{22}$—probability of observing threat occurrence in a dangerous state (*true positive*), and
   - $t(crash|deadend) = 1$—probability of observing the system crash under threat.

6. **Costs**—quantitative measures of the costs of taking actions which can be interpreted differently, depending on the needs; we are considering it to be the delay caused by the additional counteractions to neutralize the detected threats:
   a. Running cost $r(c)$ calculated as follows: $r(noact) = 0$, $r(respond) = -c$ where $c > 0$ is the penalty for executing counteraction *respond*;
   b. Final reward $R(s_F)$. This is the reward we get in the final state s_F. We define it as follows: $R(safe) = R(danger) = 1$, $R(deadend) = 0$. Thus, we consider the final state s_F as an indicator whether the transaction was successful or not.

7. **Horizon** N—length of the transaction, measured by the number of situations along the transaction.

A security decision $\phi(n)$ is a function which on each step n of the transaction chooses either *noact* or *respond*. Decision policy $\pi = (\phi(1), \phi(2), \ldots, \phi(N))$ is a collection of security decisions such that on each step n of the transaction, $\phi(n)$ depends only on the past observations, and the prior probabilities of the states before the transaction has begun. We assume that the prior probability of state *deadend* is 0, since otherwise any policy makes no sense. Therefore, the sum of the prior probabilities of the other two states is equal to 1, and the prior distribution of the states is determined by the prior probability x of state *safe*. In [4] we studied a problem with a criterion

$$v^{\pi}(x) = E_x^{\pi}(R(s_F) - cK), \tag{30}$$

where E_x^π is the expectation, corresponding to the policy π and the prior probability x, and K is the total number of times when we apply the action *respond*. A value function of the POMDP model is the function

$$v(x) = \max_\pi v^\pi(x). \tag{31}$$

The policy π such that $v(x) = v^\pi(x)$ is an *optimal policy*.

Element 6 of the model M, the running cost, is crucial. The point is that if we allow it to be arbitrary, we can face the following two extreme situations:

- **Case 1:** c is too small. Then the optimal policy may recommend applying mitigating action *respond* permanently, without taking into consideration any observations;

- **Case 2:** c is too large. Then we would not be able to neutralize all threats even if we know exactly when they occur, i.e., when both the false positives and the false negatives are equal to 0. This could happen because the total cost of the mitigating actions may exceed 1, which is the maximum reward we can get for a successful end of the transaction.

In [5], we suggested the following choice of c. Let m be the maximum number when the action *respond* can be used. Then, we have

$$mc \leq 1, \quad (m+1)c \geq 1, \tag{32}$$

and therefore,

$$\frac{1}{m+1} \leq c \leq \frac{1}{m}. \tag{33}$$

Since the number of attacks during the transaction has a Binomial distribution with parameters $1-p$ and N, a suitable choice of m is it to be $1-\alpha$-quantile of this distribution for a given significance level α.

Remark. Formula (33) assigns to c values inside the interval to which it must belong. However, in fact we should select between the end points of that interval. If we take $c = 1/(m+1)$, then we favor the successful end of the transaction. Otherwise, if $c = 1/m$, we focus on avoiding a unnecessary use of mitigating actions.

In the model described above, the probability of attacks, $1-p$, remains unchanged during the transaction. This makes the model too restrictive. Indeed, in the real life, if the computer is infected before the transaction, it is more likely to expect intrusions at its early steps rather than in the end. Moreover, some transaction steps like entering an IBAN for money transfer or credentials, are more risky than the others, purely technical, steps of the transactions. Here, we relax this restriction assuming that $p = (p_1, p_2, \ldots, p_N)$ is a vector, whose entry p_n is equal to the probability to stay in the state *safe* at the n-th step of the transaction. It is worth noting that p_1 is exactly the prior probability x in the model M, and $1-p_1$ is a probability of attacks which, however, occur not during the transaction, but before it. We shall refer to this model as a generalized model, and denote it by M'.

Let us discuss the running costs in this model. As we have seen, they are very important, since they control the actions we apply on different steps of the transaction. If the probability p_n is large, then the corresponding running cost c_n should be large as well, in order to prevent a useless choice of a mitigating action. Evidently, such an action should be applied in case of a probability p_n which is small enough, and we favor its application taking c_n to be small, too. Thus, c_n depends on p_n (but not only), and we define the running costs in this model as a vector $c = (c_1, c_2, \dots, c_N)$. The other elements of both models M and M' coincide. Eq. (30) in model M' reads as

$$v^{\pi}(x) = E_x^{\pi}\left(R(s_F) - \sum_{i=1}^{N} c_i 1(a_i = respond) \right). \tag{34}$$

We stress the dependence of c_n on p_n, making the following basic assumption.

Assumption 1 For any two steps i and j of the transaction, one holds $c_i = c_j$ if $p_i = p_j$.

This assumption trivially holds in model M, and it implies that the running cost in that model is a constant.

4.1 Determining costs $c_1, c_2, \dots, c_N$

Here, we are going to determine the costs $c_1, c_2, \dots, c_N$, trying to satisfy the following conditions:

- guarantee a successful end of the transaction with a probability which is large enough;
- avoid a useless application of the action *respond* when the current state of the transaction is *safe*.

Unlike the model M when, given N, p, and a significance level α, one should take c according to the remark after formula (33), here the situation is more complicated.

We shall consider the transaction as a random trial whose basic space Ω is the set of all diadic representation of integers between 0 and $2^N - 1$:

$$\Omega = \{\omega = w_1 w_2 \dots w_N, \quad w_n = 0 \text{ or } 1, \quad n = 1, 2, \dots, N\}.$$

We assume that $w_n = 1$ (resp. $w_n = 0$) corresponds to the event when there is (resp. there is no) an intrusion on the n-th step of the transaction. Each elementary event $\omega \in \Omega$ is determined by the set

$$A_\omega \in 2^{\{1,2,\dots,N\}}, A_\omega = \{n : \quad w_n = 1\}, \tag{35}$$

where $2^{\{1,2,\dots,N\}}$ is the set of all subsets of $\{1, 2, \dots, N\}$.

Formula (35) implies that there exists 1–1 correspondence between Ω and $2^{\{1,2,\dots,N\}}$. If necessary, we shall denote by ω_A the elementary event $\omega \in \Omega$, that corresponds to the set $A \subseteq \{1, 2, \dots, N\}$. The probability q_ω of $\omega \in \Omega$ can be calculated by the formula

$$q_\omega = \prod_{n=1}^{N} p_n^{1-w_n}(1-p_n)^{w_n}. \quad \omega \in \Omega. \tag{36}$$

Let us fix a significance level α, and let

$$q^{(1)} \le q^{(2)} \le \dots \le q^{(2^N)} \tag{37}$$

be the ordered statistics of series (q_ω), $\omega \in \Omega$. Denote by A_i the set given by (35), that corresponds to $q^{(i)}$ [1], and define the integer I by the formula:

$$I = \min\left(i : \sum_{j=1}^{i} q^{(j)} > \alpha, \text{ and } q^{(i)} < q^{(i+1)} \right). \tag{38}$$

4.1.1 Case 1

All probabilities p_n, $n = 1, 2, \dots, N$, are different.

First, we shall discuss the case when all probabilities p_n, $n = 1, 2, \dots, N$ are different. Similarly to the inequalities in (32), we want that the costs $c_1, c_2, \dots, c_N$ satisfy the constraint

$$\sum_{j=1}^{N} w_j c_j = \sum_{j \in A_i} c_j \ge 1, \omega = \omega_{A_i}, \quad i = 1, 2, \dots, I, \tag{39}$$

$$\sum_{j=1}^{N} w_j c_j = \sum_{j \in A_i} c_j \le 1, \omega = \omega_{A_i}, i = I+1, I+2, \dots, 2^N. \tag{40}$$

The mean cost of the transaction is equal to

$$C = \sum_{i=1}^{N} (1 - p_i) c_i. \tag{41}$$

In view of the remark after formula (33) is natural to try to minimize C, in order to favor the successful end of the transaction. Thus, we come to the following linear program: minimize (41), under constraint (39) and (40). Evidently, if N is large, this problem seems to become hopeless to solve even with the help of supercomputers.

4.1.2 Case 2

Some of the probabilities $p_1, p_2, \dots, p_N$, coincide.

Now, assume that some of the probabilities $p_1, p_2, \dots, p_N$ coincide. At a first glance in this case the problem becomes even more complicated, because, in view of Assumption 1, to the inequality constraint (40) and (41), one should add equality

---

[1] In fact, if the inequalities in (24) are not strong, then $q^{(i)}$ does not allow us to determine the corresponding ω, and, therefore A_i. We avoid this difficulty, defining I by formula (24).

constraint $c_i = c_j$ for all pairs of states (i,j) such that $p_i = p_j$. However, it turns out that the more equal probabilities among $p_1, p_2, \ldots, p_N$ there are, the simplier the problem becomes. Let

$$P_1 < P_2 < \ldots < P_K, \quad K < N,$$

be all different numbers among $p_1, p_2, \ldots, p_N$. Then, we can represent the set $\{1, 2, \ldots, N\}$ as a union of disjoint subsets $B_1, B_2, \ldots, B_K$ such that

$$B_k = \{n : \; p_n = P_k\}, \quad k = 1, 2, \ldots, K.$$

Assumption 1 implies that not only the probabilities (p_n), $n = 1, 2, \ldots, N$, but also the costs (c_n), $n = 1, 2, \ldots, N$ coincide on each set $B_1, B_2, \ldots, B_K$. Denote them by $C_1, C_2, \ldots, C_K$. Then, in terms of $C_1, C_2, \ldots, C_K$, both the constraint (39) and (40) as well as formula (41) reduce to

$$\sum_{k=1}^{K} d_{ik} C_k \geq 1, \quad i = 1, 2, \ldots, I, \tag{42}$$

$$\sum_{k=1}^{K} d_{ik} C_k \leq 1, i = I+1, I+2, \ldots, 2^N, \tag{43}$$

$$C = \sum_{k=1}^{K} D_k (1 - P_k) C_k, \tag{44}$$

where d_{ik}, $k = 1, 2, .., K$, $i = 1, 2, \ldots, 2^N$, and D_k, $k = 1, 2, .., K$, are cardinalities of sets $A_i \cap B_k$, and B_k, respectively. Thus, the problem becomes: minimize (44) under constraint (43) and (44). The dimensionality of this problem is K rather than N. Moreover, the number of constraint will be less than 2^N. In order to understand this, we give the following definition.

Definition 3 We say that the vector $a = (a_1, a_2, \ldots, a_K)$ majorates the vector $b = (b_1, b_2, \ldots, b_K)$, and write $a \succcurlyeq b$, if $a_k \geq b_k$, $k = 1, 2, .., K$..

Denote by Δ the matrix with entries (d_{ik}), and let $\delta_i = (d_{i1}, d_{i2}, \ldots, d_{iK})$, $i = 1, 2, \ldots, 2^N$, be its rows. If $i, j \leq I$ and $\delta_i \succcurlyeq \delta_j$, then we can discard the constraint δ_i, because it will hold automatically if we satisfy δ_j. Similarly, if $i, j > I$, and $\delta_i \succcurlyeq \delta_j$, then we can discard the constraint δ_j, because it will hold automatically if we satisfy δ_i. Thus, the number of constraint in (5) (resp. (4)) is equal to the cardinality of the maximum subset of $\{1, 2, \ldots, I\}$ (resp. $\{I+1, I+2, \ldots, 2^N\}$) such that for any two i and j belonging to this subset neither $\delta_i \succcurlyeq \delta_j$ nor $\delta_j \succcurlyeq \delta_i$.

4.2 An example

A transaction consists of $N = 12$ steps with probabilities

$$p = (0.6, 0.9, 0.9, 0.8, 0.7, 0.9, 0.9, 0.9, 0.9, 0.9, 0.8, 0.9).$$

We have $K = 4$, $P = (0.6, 0.7, 0.8, 0.9)$, $D = (1, 1, 2, 8)$. The linear program (42), (43) reduces to: minimize

$$0.4C_1 + 0.3C_2 + 0.4C_3 + 0.8C_4,$$

under constraint

$$\begin{aligned} C_1 + 3C_4 &\geq 1, \\ C_2 + 3C_4 &\geq 1, \\ C_3 + 3C_4 &\geq 1, \\ C_3 + C_4 &\geq 1/2, \\ C_2 + C_3 + 2C_4 &\geq 1, \\ C_1 + C_3 + 2C_4 &\leq 1, \\ C_1 + C_2 + 2C_4 &\leq 1, \\ C_1 + C_2 + 2C_3 + C_4 &\leq 1, \end{aligned}$$

where

$$0 \leq C_1 \leq 1, \;\; 0 \leq C_2 \leq 1, \;\; 0 \leq C_3 \leq 1/2, C_4 = 1/3.$$

Its solution is $C_1 = 0, \;\; C_2 = C_3 = 1/6, \;\; C_4 = 1/3$. Thus, in the first state of the transaction, which is most dangerous, one should pay a zero cost for counteracting. Therefore, we shall certainly prevent an attack in this state, if it occurs.

4.3 Description of the optimal policy

Here, we briefly describe the optimal policy in model M', referring the reader for more details to [4], where, in case of the model M, both the value function (31) and the optimal policy have been found by performing a standard reduction to a fully observable MDP. Sufficient statistics in this reduction are the posterior probabilities of the states of the transaction. Since the state *deadend* is absorbing, the total probability of the other two states, *safe* and *danger*, is 1, and we can consider the posterior probability of either of them, say *safe*, as a controlled process in the reduced MDP model. The state space of that process is $[0, 1] \cup \{*\}$, where the isolated point $*$ corresponds to state *deadend*. Consider the following functions defined for $x \in [0, 1]$:

$$\Gamma^1(x, 1-x) = \frac{p_{21}x}{p_{21}x + p_{22}(1-x)}, \Gamma^2(x, 1-x) = \frac{p_{11}x}{p_{11}x + p_{12}(1-x)}, \tag{45}$$

$$V_n(x) = \max_\pi E_x^\pi \left(1(0 \leq x_{N+1} \leq 1) - \Sigma_{k=n}^N c_k 1(a_k = \textit{respond})\right), n = 1, \ldots N, \tag{46}$$

where $x = x_n$ is the posterior probability of state *safe* at time n. The function $\Gamma^1(x, 1-x)$ (resp. $\Gamma^2(x, 1-x)$) in (45) is the posterior probability of state *safe*, if we detect a *danger* (resp. *nothreat*), and the prior probability of state *safe* is x. The functions in (46) represent the maximum income we can get after the n-th step of the transaction. They satisfy Bellman's equation

$$V_n(x) = \max\left(V'_n(x), V''_n(x)\right), \tag{47}$$

and the final condition

$$V_{N+1}(x) = 1. \tag{48}$$

In (47), $V'_n(x)$ and $V''_n(x)$ are one-step ahead estimates of both actions *noact* and *respond*:

$$\begin{aligned} V'_n(x) &= x(p_n p_{21} + (1-p_n)p_{22})V_{n+1}(\Gamma^1(p_n, 1-p_n)) \\ &+ x(p_n p_{11} + (1-p_n)p_{12})V_{n+1}(\Gamma^2(p_n, 1-p_n)), \\ V''_n(x) &= -c_n + (p_n p_{21} + (1-p_n)p_{22})V_{n+1}(\Gamma^1(p_n, 1-p_n)) \\ &+ (p_n p_{11} + (1-p_n)p_{12})V_{n+1}(\Gamma^2(p_n, 1-p_n)). \end{aligned}$$

The optimal strategy φ_n at any moment of time $n = 1, 2, \ldots, N$ is the following:

$$\varphi_n(x) = \begin{cases} noact, & \text{if } V_n(x) = V'_n(x) \\ respond, & \text{if } V_n(x) = V''_n(x) \end{cases}.$$

Eqs. (47) can be solved backwards, starting with the boundary condition (48). For example, for $n = N$ we get:

$$V_N(x) = \max(1 - c_N, x),$$

$$\varphi_N(x) = \begin{cases} noact, & \text{if } x \geq 1 - c_N \text{ (above the threshold)} \\ respond, & \text{if } x < 1 - c_N \text{ (bellow the threshold)} \end{cases}.$$

The remaining iterations until reaching the beginning of the transaction can be performed recursively, taking the previously calculated solution as terminal. It is worth noting that not only φ_N, but also all $\varphi_n, n = 1, 2 \ldots, N-1$ are determined by thresholds $(y_n), n = 1, 2, \ldots, N-1,\ y_N = 1 - c_N$, and the optimal policy recommends using of a counteraction only if the posterior probability of state *safe* in time $n,\ x_n,$ is less than y_n.

After all thresholds have been found, the optimal policy recommends the following optimal behavior:

1. At the beginning of transaction ($n = 1$), depending on whether we detect a threat or not, we calculate the posterior probability x_1 of the state *safe* by the formula

$$x_1 = \begin{cases} \Gamma^1(p_1, 1-p_1), & \textit{if } z_1 = \textit{threat} \\ \Gamma^2(p_1, 1-p_1), & \textit{if } z_1 = \textit{nothreat} \end{cases},$$

   If x_1 is greater than the threshold y_1, then we do not apply the counteraction, x_1 remains unchanged, and with probability $1 - x_1$ we fall into a state *deadend*. Otherwise, we use *respond* after paying the cost c_1, and the posterior probability becomes 1 since we certainly know that we are *safe*;

2. On the next step we start with a prior probability equal to the just found posterior probability (x_1 or 1), multiplied by p_2, observe again if there is detection of a threat, calculate the posterior probability x_2, compare it with the threshold y_2, and so on till the end of the transaction.

5. Conclusions

We focus on two new applications of the theory of the Partially Observable Markov Decision Processes (POMDPs).

Firstly, we revisit the discrete time disorder problem. The classical approach to it is based on optimal stopping rules and martingale technique. We include it into the framework of POMDPs which improves the quality of detection, and allows, in some cases, to find solutions to Bellman's equation.

The second application is in the cybersecurity. A basic problem here is the assessment of the security risk during transaction processing. We consider the transaction as a controlled POMDP and construct an optimal policy that recommends a best behavior for neutralizing the intrusions when they occur. We reveal the crucial role of the costs of counteractions in building the model. On the one hand, they should favor the successful end of the transaction, but, on the other hand, prevent the unnecessary use of counteractions when there is no an attack. We show that the costs can be found as optimal solutions of some linear program.

Abbreviations

POMDP	Partially Observable Markov Decision Processes
MDP	Markov Decision Processes
IDS	Intrusion Detection Systems
ML	Machine Learning

Author details

Doncho S. Donchev
Sofia University "St. Kliment Ohridski", Sofia, Bulgaria

*Address all correspondence to: d.donchev@fmi.uni-sofia.bg

References

[1] Shiryev A. Optimal Stopping Rules. Berlin, Heidelberg: Springer; 2008

[2] Kulariya M et al. Performance analysis of network intrusion detection schemes using apache spark. In: Proc. 2016 Int. Conf. On Communication and Signal Processing (ICCSP). Melmaruvathur: IEEE; 2016. pp. 1973-1977

[3] Parmar JA. Classification based approach to create database policy for intrusion detection and respond anomaly requests. In: Proc. 2014 Conf. On IT in Business, Industry and Government (CSIBIG). Indore: IEEE; 2014. pp. 1-7

[4] Donchev D, Tonchev D, Vassilev V. Risk assessment in transactions under threat as a partially observable Markov decision process. In: Dellolmo P et al., editors. Optimization in Artificial Inteligence and Data Science, Proc. of Int. Conference and Decision Science ODS2021, 14–17 Sep, 2021. Rome, Italy: Springer; 2021. pp. 199-212

[5] Donchev D, Tonchev D, Vassilev V. Impact of false positives and false negatives on security risks in transactions under threat. In: Fischer-Hubner S, Lambrinoudakis C, Kostis G, et al., editors. Trust, Privacy and Security in Digital Business, LNCS. Vol. 12927. Springer; 2021. pp. 50-68

[6] Bertsekas D, Shreve S. Stochastic Optimal Control. The Discrete Time Case. New York, San Francisko, London: Academic Press; 1978

[7] Dynkin E, Yushkevich A. Controlled Markov Processes. New York: Springer; 1979

[8] Brychkov U, Marychev O, Prudnikov A. Integrals and Series.Vol. 3. Moscow: Nauka; 1986

[9] Donchev D. Random series with time-varying discounting. Communications in Statistics-Theory and Methods. 2011; **40**(16):2866-2878

Chapter 6

Engineering Stochastic Optical Modes

Patricia Martinez Vara, Juan Carlos Atenco Cuautle, Elizabeth Saldivia Gomez and Gabriel Martinez Niconoff

Abstract

In this chapter, we describe the engineering of optical modes whose axial structure follows fluctuations of Markov-chain type. The stochastic processes are associated with a sequence of time subintervals of duration ΔT. Each subinterval is linked to a Bessel mode of integer order. This process models a thermodynamic equilibrium and can be related to the evolution and stability of optical systems. The matrix representation for the stochastic process allows the incorporation of entropy properties and therefore, it is possible to deduce the similarity with completely coherent modes. This property is known as the purity of the optical mode. Herein, the resulting optical field is simulated using Markov-chain type Ehrenfest process.

Keywords: Markov chain, Ehrenfest process, stochastic process, entropy, N-step matrix

1. Introduction

The time/spatial evolution of the optical fields is determined by its coherence features, which are closely related to the fluctuations of polarization states [1]. In this context, coherence deals with the propagation of the statistical correlation function. The control of the coherence properties has been implemented in the generation of tunable optical tweezers [2], which allows the generation of novel spectroscopy effects described by inducing resonant effects that locally modify the effective refractive index [3]. In the context of communications systems, it enables the transmission of a great volume of information [4]. From the theoretical point of view, the coherence study is interesting because of phase dislocation at the boundary conditions that can be matched with the generation of topological charge. By giving it a dynamic behavior, it is possible to induce topological currents and consequently, the media acquire a topological conductivity with the corresponding entropy fluctuations. However, the coherence models have been designed under the hypothesis that the random fluctuations of the optical field have a stationary character [5, 6]. For the study of the partial coherence effects, some parameters that characterize the balance among the amplitude-phase functions are time dependent [7]. Therefore, the topological charge acquires a dynamical behavior that justifies the generation of topological currents and consequently, topological conductivities emerge. This behavior implies a strong analogy to the traditional electromagnetic theoretical models. In this chapter, we establish the basis to describe the generation of optical fields that exhibits a random behavior without the stationary hypothesis. Then, we perform the synthesis of optical fields with time-dependent partially coherent features whose evolution follows a Markov-chain type Ehrenfest process. Using this type of chain allows the establishment of a strong analogy between the evolution of optical fields and the thermodynamic models in equilibrium. The equilibrium behavior is known because stochastic optical fields increase their coherence properties as they propagate. The theoretical predictions are supported by computational simulations, where the coherence parameters are related to the entropy process allowing the definition of a purity factor for the optical fields. This factor is complementary to the degree of coherence, which allows us to identify and compare the similarity with a fully coherent optical field.

2. Brief overview of the coherence models

We start our analysis with an overview of the coherence features. Coherence is the analysis of the correlation complex amplitude function between two arbitrary points in time/space. This study is supported by assuming that the optical fields under study are stationary and ergodic. The simplest case occurs when we describe the autocorrelation functions [8] defined as

$$< \phi(t_1)\phi^*(t_2) > = \int \phi(t_1)\phi^*(t_2)\rho(t_1,t_2)dt_1dt_2, \tag{1}$$

in most cases, the probability density function $\rho(t_1,t_2)$ is an unknown function. When the random fluctuations are stationary, the probability density function depends on the time difference, and it is of the form

$$\rho(t_1,t_2) = \rho(t_1 - t_2). \tag{2}$$

As consequence, the autocorrelation function acquires the form of a convolution integral, simplifying the calculus of the coherence function. To describe the amplitude distribution of the optical field, we use the fact that we can represent it as a sum of modes, given by

$$\phi(x,t) = \sum a_n(t)e^{i\beta z}f_n(x,y), \tag{3}$$

Where $f_n(x,y)$ are functions that describe the mode profile, and its square modulus gives its morphological shape. The coefficients are random time-depending parameters. The amplitude product for two arbitrary optical fields takes the form

$$\phi(x,t_1)\phi^*(x,t_2) = \frac{1}{nm}\sum_n\sum_m a_n(t_1)a_n^*(t_2)f_n(x,y)f_m^*(x,y), \tag{4}$$

And the autocorrelation function defined by the mean of the previous equation takes the form

$$C(t_1,t_2) = < \phi(x,t_1)\phi^*(x,t_2) > = \frac{1}{nm}\sum_n\sum_m < a_n(t_1)a_n^*(t_2) > f_n(x,y)f_m^*(x,y). \tag{5}$$

from the previous representation, we can easily deduce the matrix structure for the autocorrelation function. Rewriting this correlation function in a matrix form as

$$C(t_1,t_2) = < \frac{1}{nm}(f_0\, f_1\, \cdots\, f_n), \begin{pmatrix} a_0(t_1)a_0^*(t_2) & a_0(t_1)a_1^*(t_2) & \cdots & a_0(t_1)a_m^*(t_2) \\ a_1(t_1)a_0^*(t_2) & a_1(t_1)a_1^*(t_2) & \cdots & a_1(t_1)a_m^*(t_2) \\ \vdots & \vdots & \ddots & \vdots \\ a_n(t_1)a_0^*(t_2) & a_n(t_1)a_1^*(t_2) & \cdots & a_n(t_1)a_m^*(t_2) \end{pmatrix} \begin{pmatrix} f_0^* \\ f_1^* \\ \vdots \\ f_m^* \end{pmatrix} >, \tag{6}$$

Where the square brackets mean the expected value is evident that the correlation matrix is Hermitian. Each term in the matrix carries on information about the interaction between all possible modes.

The matrix representation allows the implementation of a Markovian process where $<a_p(t_1)a_q^*(t_2)>$ is closely related to a transition probability. Modifying the matrix elements is possible to generate optical fields that do not present a stationary behavior. In particular, we focus our attention on the generation of optical fields whose random fluctuations follow a Markov chain.

3. Markovian processes: Some basic concepts

In many physical situations, we are interested in describing the occurrence of a set of random events represented by $P(\mu_0, \ldots, \mu_n)$, where μ_i is linked to a certain physical situation. This expression can be rewritten using Bayes formula as

$$P(\mu_0, \mu_1, \ldots, \mu_n) = P(\mu_n|x_{n-1}, \ldots, \mu_1)P(\mu_{n-1}, \mu_{n-2}, \ldots, \mu_1, \mu_0), \tag{7}$$

the previous expression can be simplified assuming that its behavior depends only on its recent past [9], i.e., $P(\mu_n|\mu_{n-1}, \ldots, x_1) = P(\mu_n|\mu_{n-1})$. Applying recursively this expression, Eq. (7) acquires the form

$$P(\mu_0, \mu_1, \ldots, \mu_n) = P(\mu_n|\mu_{n-1})P(\mu_{n-1}|\mu_{n-2}) \ldots P(\mu_{n-2}, \ldots, \mu_1) \ldots P(\mu_1|\mu_0)P(\mu_0). \tag{8}$$

Eq. (8) can be interpreted as the n-th correlation order, in this way, we can expect that this expression can be used for the analysis of optical coherence. Eq. (8) defines the Markov chain, where the term $P(\mu_i|\mu_{i-1})$ is known as the transition probability. With the transition probabilities, it is possible to generate a matrix representation given by

$$\mathbf{P}(\mu_i|\mu_j) = \begin{pmatrix} \mu_{00} & \cdots & \mu_{0n} \\ \vdots & \ddots & \vdots \\ \mu_{n0} & \cdots & \mu_{nn} \end{pmatrix}, \tag{9}$$

where the matrix elements correspond to the transition probabilities. This expression is known as the transition matrix. As each matrix term represents a probability, the sum of the row elements satisfies the conservation probability given by

$$\sum_{j} \mu_{ij} = 1, \quad i = 0,1,2, \ldots, n. \tag{10}$$

This means that the Markov chain has associated with a stochastic matrix representation. Important properties of a stochastical matrix can be found in Ref. [10].

The evolution of a Markov chain can be obtained by applying the transition matrix to a random vector, which represents the probability of the chain starting in a given situation. Then, the stochastic matrix is acting on a random vector transforming it into another random vector. This representation is given by

$$\vec{\alpha}_1 = \begin{pmatrix} a_0 & \cdots & a_n \end{pmatrix} \begin{pmatrix} \mu_{00} & \mu_{01} & \cdots & \mu_{0n} \\ \mu_{10} & \mu_{11} & \cdots & \mu_{1n} \\ \vdots & \vdots & \ddots & \vdots \\ \mu_{n0} & \mu_{n1} & \cdots & \mu_{nn} \end{pmatrix} = \vec{\alpha}_0 \mathbf{P}, \tag{11}$$

and $\alpha_2 = \alpha_1 \mathbf{P}^2 = \alpha_0 \mathbf{P}^2, \ldots, \alpha_n = \alpha_0 \mathbf{P}^n$ where

$$\mathbf{P}^n = \begin{pmatrix} \mu_{00} & \cdots & \mu_{0n} \\ \vdots & \ddots & \vdots \\ \mu_{n0} & \cdots & \mu_{nn} \end{pmatrix}^n, \tag{12}$$

is the n-step transition matrix.

4. Markov chain-type Ehrenfest process

In this section, we analyze the equilibrium of a Markovian process that follows a Markov chain-type Ehrenfest process [11]. This is performed by using a boxes model. The simplest case occurs by considering two boxes labeled as B_1 and B_2. In the boxes are randomly distributed n balls. In box B_1 are s balls, and in the box B_2 are $n-s$ balls. The balls are labeled from 1 to n and they are randomly distributed in each box. The Ehrenfest process describes the transfer of the balls between the boxes. The first step is the random selection of a ball and transferring it to the other box with probability β, or letting it to remain in the same box with probability $1-\beta$. This process is repeated many times and the resulting ball distribution acquires a stable configuration. It can be shown that the Markovian matrix is given by

$$E = \begin{pmatrix} (1-\beta) & \beta & 0 & 0 & 0 & \cdots & 0 \\ \frac{\beta}{n} & (1-\beta) & \left(\frac{n-1}{n}\right)\beta & 0 & 0 & \cdots & 0 \\ 0 & \frac{2\beta}{n} & (1-\beta) & \left(\frac{n-2}{n}\right)\beta & 0 & \cdots & 0 \\ 0 & 0 & \frac{3\beta}{n} & (1-\beta) & \left(\frac{n-3}{n}\right)\beta & \cdots & 0 \\ \vdots & \vdots & \vdots & \vdots & \vdots & \ddots & \vdots \\ 0 & 0 & 0 & 0 & \cdots & \beta & (1-\beta) \end{pmatrix}. \tag{13}$$

DOI: http://dx.doi.org/10.5772/intechopen.1002038

To analyze the convergence and the stability of the Ehrenfest process, we implement a numerical example for four balls and $\beta = 1/2$. We focus our attention on the following issues: how the probabilistic terms in the rows are transformed and how the matrix evolves after N steps. The following matrix expressions show the resulting stochastic matrix after 1, 2, and 25 steps.

$$E = \begin{pmatrix} 0.5000 & 0.5000 & 0 & 0 \\ 0.1250 & 0.5000 & 0.3750 & 0 \\ 0 & 0.2500 & 0.5000 & 0.2500 \\ 0 & 0 & 0.5000 & 0.5000 \end{pmatrix}, \quad E^2 = \begin{pmatrix} 0.3125 & 0.5000 & 0.1875 & 0 \\ 0.1250 & 0.4063 & 0.3750 & 0.0938 \\ 0.0313 & 0.2500 & 0.4688 & 0.2500 \\ 0 & 0.1250 & 0.5000 & 0.3750 \end{pmatrix},$$
$$\cdots, \quad E^{25} = \begin{pmatrix} 0.0714 & 0.2857 & 0.4286 & 0.2143 \\ 0.0714 & 0.2857 & 0.4286 & 0.2143 \\ 0.0714 & 0.2857 & 0.4286 & 0.2143 \\ 0.0714 & 0.2857 & 0.4286 & 0.2143 \end{pmatrix}. \tag{14}$$

From the last matrix expression, we can identify that all elements in each column have the same value; consequently, all rows have the same entropy value that corresponds to the final equilibrium state [12]. The structure of Eq. (14) shows that the product of a row in the arbitrary random vector with an N-step stochastic matrix reproduces the matrix row. This represents the equilibrium of the process, and a surprising result is that the final entropy does not depend on the initial random vector. From this entropy value, it is possible to deduce some generic features. For example, the maximum entanglement between the probabilistic values in the n-step random matrix. This can be obtained by analyzing the digraphs evolution shown in **Figure 1**.

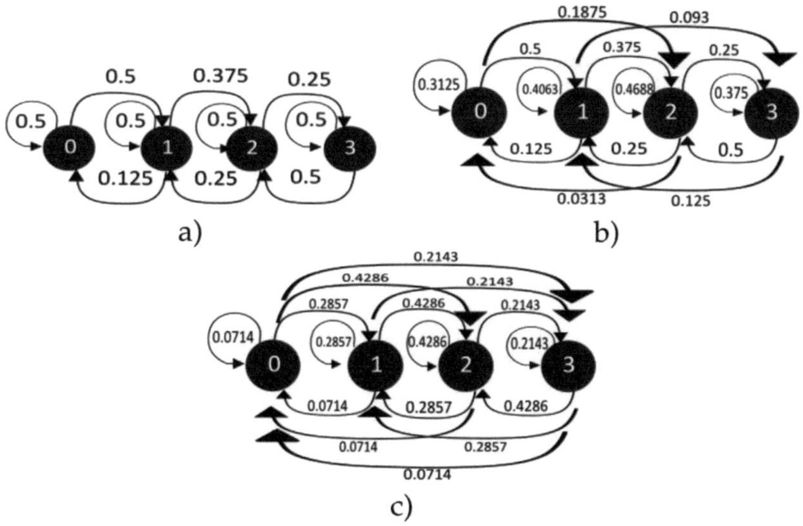

Figure 1.
Depicture of the stochastic matrix process corresponding to Eq. (14). (a) Digraph associated with the initial stochastic matrix. (b) Digraph for E^2. (c) Digraph for E^{25}, showing when the process reaches a standard configuration.

	0	1	2	3
0	0.0714	0.2857	0.4286	0.2143
1	0.0714	0.2857	0.4286	0.2143
2	0.0714	0.2857	0.4286	0.2143
3	0.0714	0.2857	0.4286	0.2143

Table 1.
*Numerical representation of the digraph shown in **Figure 1(c)**.*

The entanglement features carry information on the evolution of the different probability states between the nodes. This analysis is contained in the entropy value evolution.

Finally, we remark that the changes in the assigned probability values in the initial stochastic matrix imply the modification of the connectivity between states. This information becomes evident in **Table 1**; however, all the nodes acquire invariant connectivity between the probabilistic states obtained from the n-step probabilistic matrix.

5. Generation of Ehrenfest optical modes

The purpose of this section is to implement an Markovian Ehrenfest process in the optical context. To perform this, we define an optical mode as an exact solution to the Helmholtz equation. The mathematical expression is of the form [13]:

$$\phi(x,y,z) = f(x,y)\exp(i\beta z), \tag{15}$$

where the function $f(x, y)$ must satisfy the eigenvalue equation

$$\nabla_{\perp}^{2} f(x,y) + K^{2} f(x,y) = \beta^{2} f(x,y), \tag{16}$$

the transversal profile $f(x, y)$ remains invariant when it is propagating along the z coordinate. For mathematical convenience, we solve Eq. (16) by using polar coordinates, allowing us to identify the solutions as a set of Bessel functions of integer order. The optical modes are given by

$$\left\{e^{i\beta z} J_n(2\pi r d) e^{in\theta}\right\} \quad n = 0, \pm 1, \pm 2, \dots . \tag{17}$$

It should be noted that all of the modes have the same phase value that describes the mode propagation along the z coordinate. This is a condition for all the modes present diffraction-free features. We will use the representation to generate a tandem array of optical modes following a Markovian chain, i.e., we engineer a stochastical mode that locally presents diffraction-free features. An important case occurs when a set of integer-order Bessel modes are selected following a Markov chain type process.

Eq. (17) can be matched with the box model of the Ehrenfest process described in the previous section. By replacing the label in each ball by $J_0, J_1, \dots, J_n$ the evolution of initial state is described $x_0 = (a_0, \dots, a_n)$. This vector corresponds to the coordinates representing the appearance of the mode with the following interpretation: Assuming that the process has time duration T, divided by n subintervals of length ΔT. In each subinterval, a Bessel mode of integer order is selected. Thus, the optical field consists of a succession of mode-type chains where the occurrence of the ith Bessel mode is $n\alpha_i$. With this interpretation, the optical field assumes the structure shown in **Figure 2**. A liquid crystal display (LCD) is implemented to generate the boundary condition that consists of an annular slit angularly modulated for synthesizing the corresponding Bessel mode [13, 14]. The chain structure corresponds to an Ehrenfest mode.

To get an understanding of the evolution of the Markovian mode, it is convenient to describe a tree graph, shown in **Figure 3**. The dotted line represents the sequence

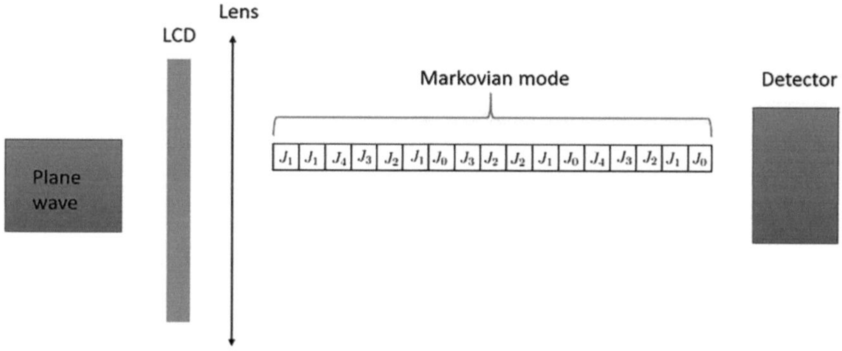

Figure 2.
To generate the Bessel modes, an LCD is illuminated containing an annular slit with time-dependent angular modulation with a coherent plane wave linearly polarized.

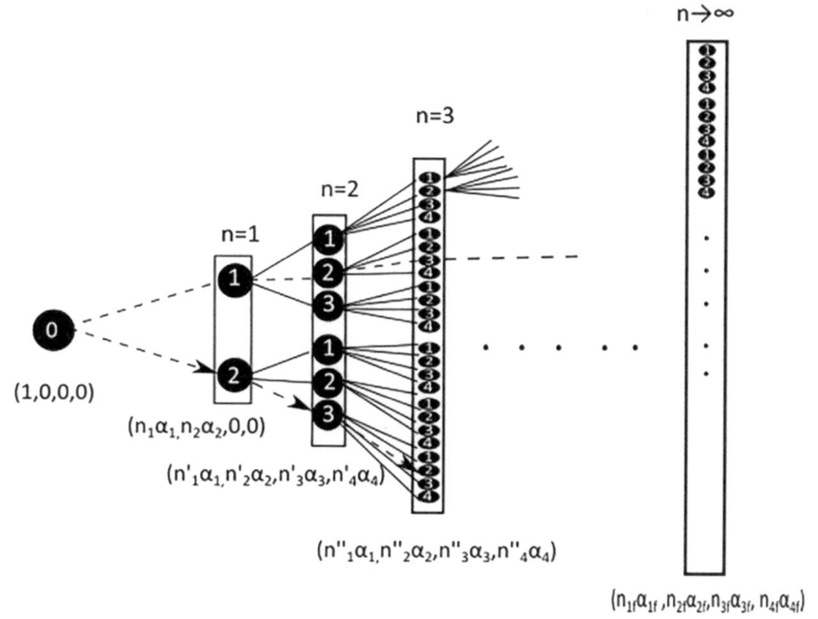

Figure 3.
Sketch of the entanglement between states as "n" grows; during this evolution, it is possible to visualize that the probability density is distributed between the accessible states.

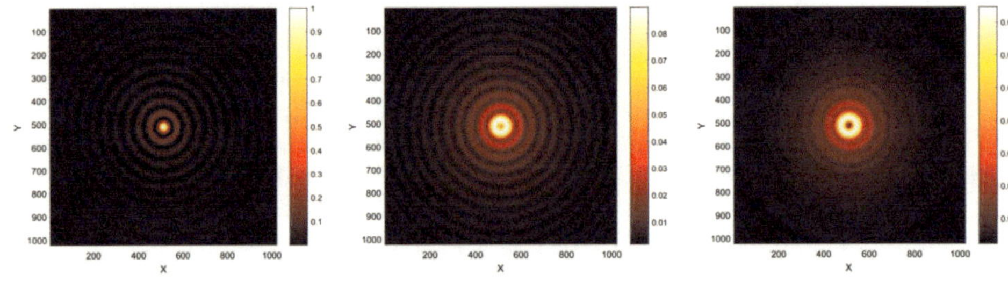

a) b) c)

Figure 4.
The mean irradiance after N-steps for an Ehrenfest-type process. (a) Represents the initial state associated with a J_0 Bessel mode. The initial probability vector is (1, 0, 0, 0). (b) Mean irradiance after 2-steps, and the probability vector is (1/2, 1/2, 0, 0). (c) Mean irradiance after 25-steps, the probability vector is (0.0714, 0.2857, 0.4286, 0.2143). It must be noted that this last vector corresponds to a row for the stabilized N-step stochastic matrix. The expression for the mean irradiance can be related to the probability vector as $I={}^{(N)}n\left(P_0^2 J_0^2 + P_1^2 J_1^2 + P_2^2 J_2^2\right)$, where ${}^{(N)}nP_i$ is the occurrence number for the irradiance of each mode.

$(J_0 - J_0 - J_1 - J_2 \ldots)$, and the dotted arrowed line represents $(J_0 - J_1 - J_2 - J_1 \ldots)$. From this representation, time-structured modes are generated, and all of them must exhibit the same irradiance mean when the equilibrium is reached. In the early steps, the corresponding modes display different irradiance values as shown in **Figure 4** obtained with commercial software.

The Ehrenfest mode consists of a sequence of Bessel modes of integer order where each sequence appears according to a certain probability value. The state starts with a zero-order Bessel beam, whose irradiance is shown in **Figure 4-a**, which evolves following the chain and after three steps, the initial vector evolves toward the vector whose irradiance is given by $\alpha_0^{(2)} J_0^2 + \alpha_1^{(2)} J_1^2 + \alpha_2^{(2)} J_2^2$. When the experiment is performed n-times, the number of occurrences of each element of the basis can be obtained from the transformed random vector. The mean of irradiance after three steps is shown in **Figure 4-b**. When the number of steps increases, the chain is stabilized and the irradiance means distribution is shown in **Figure 4-c**. The global optical fields were analyzed once the equilibrium was reached. The mean irradiance distribution presents diffraction-free features. The latter is easily understood since the irradiance associated with each mode is non-depending on the z-coordinate. However, locally these fields follow a sequence of optical modes according to the Markov chain selected. The resulting optical mode consists of a sequence of time-changing blocks that do not follow a stationary process. For this reason, the mode is characterized using the entropy models in the following section.

6. Entropy, purity, and interference between Markovian modes

Using the fact that a Markovian process is associated with a stochastic matrix, some generic properties through entropy description can be identified allowing the study of the mode's structural features. The calculus of the Von Neumann entropy is proposed [15] in order to obtain a reference to compare the entropy value from the N-step stochastic matrix. The entropy is calculated from the principal diagonal elements, and we remark that the resulting value acts as a reference value for the different entropy measurements obtained from the elements of the rows in the N-steps resulting matrix; also as the entropy obtained in the secondary diagonal, this last value contains information about the correlation among the constitutive modes [16]. By comparing these entropy values, it can be deduced how the correlation function evolves, which allows us to understand the irradiance distribution as a function of N, which represents the number of applications of the initial stochastic matrix. The Von Neumann entropy is defined as

$$S_c = -\mathrm{Tr}D\left(E^N LnE^N\right), \tag{18}$$

where Tr denotes the trace of stochastic matrix $E^N\,\mathrm{Ln}\left(E^N\right)$. It should be noted that this type of entropy contains information for the irradiance distribution. However, we need to describe the entanglement among the elements of the basis. This description can be obtained by proposing the correlation entropy calculus as an ansatz using the elements in the secondary diagonal. This entropy takes the following form:

$$S_v = -\mathrm{Tr}\left(E^N\,\mathrm{Ln}E^N\right), \tag{19}$$

where TrD denotes the trace of the secondary diagonal. Furthermore, a good method for describing the mode structure is applied by calculating the difference of the entropy values, which is expressed as

$$\Delta S = S_c - S_v. \tag{20}$$

From this last definition, it can be easily proved that the correlation entropy is always lower-bounded by the Von Neumann entropy S_c, ... S_v. To compare the entropy values, each diagonal needs to satisfy the normalization condition. From Eq. (20), certain interesting cases can be identified. The limit case occurs when $\Delta S = 0$. Its physical meaning is that all irradiance events involved during the process participate in the global irradiance distribution, another case occurs when $\Delta S = S$. There is no interaction among the elements of the basis; thus, they are statistically independent. However, this case is not permitted in the Markov chain-type Ehrenfest process. Finally, another entropy measurement can be obtained from the qth-row elements, expressed as

$${}_qS_f = -\sum_i \alpha_{iq} Ln\left(\alpha_{iq}\right), \tag{21}$$

where α_{iq} denotes the elements in the qth row and satisfies $\sum_i \alpha_{iq} = 1$ for $q =$ 0,1,2, ... , n. From this set of entropy values, an order relationship is easily identified, e.g.,

$$S_q < S_2 = S_4 < S_5 < \dots, \tag{22}$$

this means that a qth-Bessel mode appears in a principal manner, followed by the second-order Bessel mode that appears in the same proportion as the fourth-order Bessel modes, and so on. From this order relationship, we can associate a purity measurement to the Ehrenfest mode [17] as follows:

$$P_q = 1 - \frac{S_q^{(n)}}{\sum_{l=0}^{N} S_i^{(n)}}, \tag{23}$$

which determines the similitude of the Markovian mode with the J_q mode because $\sum_{q=0}^{n} P_q = 1$. The entropy values for the Ehrenfest mode are given by $S_v = S_c = 0.5383$. The equality indicates that the process reaches an equilibrium condition. In addition, all of the rows have the same entropy value. Consequently, from the purity definition, it can be deduced that the resulting mode is the same for each element of the basis. This result is expected because the Ehrenfest process describes the conditions to reach an equilibrium system. Consequently, once the equilibrium condition is reached and as a result of the isotropy of the process, all the elements of the basis in the resulting random vector appear in the same proportion. The purity concept describes the times each element appears in a given mode.

7. Conclusions

We described the synthesis of Markovian optical modes that have diffraction-free features. The type of optical field was obtained by means of a tandem array or chain of Bessel modes. The chain evolves following a Markovian process. We present an example for a Markov-chain process type Ehrenfest. This type of process is very important since the theoretical point of view because it is close related to the thermodynamical process establishing a geometrical point of view for the entropy evolution. The optical mode reaches an equilibrium condition, which can be deduced from the N-step stochastical matrix. This condition is identified when the random rows of the matrix have the same entropy value. Using the entropy values we define the purity of the mode, and this definition allows to compare the similitude of the Markovian mode with a single Bessel mode. A very important result of the chapter is that changing the probabilistic values in the Markovian matrix all the Ehrenfest matrixes reaches the same final configuration; however, the number of steps is different. This property corresponds to hysteresis effect. These features allow to implement inverse processes in particular offering the possibility to generate Markovian holographic process. In the optical context, the Markovian optical mode can be used as illumination beam to generate random diffraction also as it can be implemented to generate tunable optical tweezers inducing tunable spectroscopy features [18–21]. Other immediate applications are cryptographic transference information, entanglement of arbitrary optical fields, and self-healing analysis [22–27].

Author details

Patricia Martinez Vara[1†], Juan Carlos Atenco Cuautle[2†], Elizabeth Saldivia Gomez[2†] and Gabriel Martinez Niconoff[1,2*†]

1 Facultad de Ingieneria, Benemerita Universidad Autonoma de Puebla, Puebla, Mexico

2 Departamento de Optica, Instituto Nacional de Astrofisica Optica y Electronica, Puebla, Mexico

*Address all correspondence to: gmartin@inaoep.mx

† These authors contributed equally.

References

[1] Kanseri B, Singh HK. Development and characterization of a source having tunable partial spatial coherence and polarization features. Optik. 2020;**206**: 163747

[2] Tang X, Xu X, Yan Z. Tunable optical tweezers by dynamically sculpting the phase profiles of light. Applied Physics Express. 2021;**14**:022009

[3] Shoji T, Tsuboi Y. Plasmonic optical tweezers toward molecular manipulation: Tailoring Plasmonic nanostructure, light source, and resonant trapping. Journal of Physical Chemistry Letters. 2014;**5**:2957-2967

[4] Li A et al. Enabling Technology in High-Baud-Rate Coherent Optical Communication Systems. IEEE Access. 2020;**8**:111318-111329

[5] Korotkova O, Wolf E. Changes in the state of polarization of a random electromagnetic beam on propagation. Optics Communications. 2005;**246**: 35-43

[6] Tervo J, Setl T, Friberg AT. Theory of partially coherent electromagnetic fields in the spacefrequency domain. Journal of the Optical Society of America A. 2004; **21**:2205-2215

[7] Gbur G. Partially coherent beam propagation in atmospheric turbulence. Journal of the Optical Society of America A. 2014;**31**: 2038-2045

[8] Mandel L, Wolf E. Optical Coherence and Quantum Optics. New York, NY, USA: Cambridge University Press; 1995

[9] Hoel SPH, Stone C. Introduction to Sthochastic Processes. Boston, USA: Houghton Mifflin; 1972

[10] Coleman R. Stochastic Processes, Problem Solvers. Netherlands: Springer; 1974

[11] Costantini D, Garibaldi U. The ehrenfest fleas: From model to theory. Synthese. 2004;**139**:107142

[12] Chen Y-P. Which design is better? Ehrenfest urn versus biased coin. Advances in Applied Probability. 2000; **32**:738749

[13] Durnin J. Exact solutions for nondiffracting beams. i. the scalar theory. Journal of the Optical Society of America A. 1987;**4**:651654

[14] Martinez-Niconoff G, Martinez-Vara P, Andres-Zarate E, Silva-Barranco J, Munoz-Lopez J. Synthesis of sources with Markovian features. Journal of the European Optical Society Rapid Publications. 2013;**8**:13005(1-7)

[15] Barakat R, Brosseau C. Von neumann entropy of n interacting pencils of radiation. Journal of the Optical Society of America A. 1993;**10**: 529532

[16] Selvamuthu D, Di Crescenzo A, Giorno V, Nobile A. A continuous-time ehrenfest model with catastrophes and its jump-diffusion approximation. Journal of Statistical Physics. 2015;**161**: 326345

[17] Picozzi A. Entropy and degree of polarization for nonlinear optical waves. Optics Letters. 2004;**29**: 16531655

[18] Pang Y, Gordon R. Optical trapping of a single protein. Nano Letters. 2012; **12**:402-406

[19] Cao T, Qiu Y. Lateral sorting of chiral nanoparticles using Fano-enhanced chiral force in visible region. Nanoscale. 2018;**10**:566-574

[20] Hester B, Campbell GK, Lpez-Mariscal C, Filgueira CL, Huschka R, Halas NJ, et al. Tunable optical tweezers for wavelength-dependent measurements. The Review of Scientific Instruments. 2012;**83**:043114

[21] Teeka C, Jalil MA, Yupapin PP, Ali J. Novel tunable dynamic tweezers using dark-bright soliton collision control in an optical add/drop filter. IEEE Transactions on Nanobioscience. 2010;**9**:258-262

[22] Sapozhnikov O. An exact solution to the helmholtz equation for a quasi-gaussian beam in the form of a superposition of two sources and sinks with complex coordinates. Acoustical Physics. 2012;**58**:4147

[23] Kotlyar VV, Kovalev AA, Soifer VA. Asymmetric bessel modes. Optics Letters. 2014;**39**:23952398

[24] Barnett SM, Phoenix SJD. Entropy as a measure of quantum optical correlation. Physical Review A. 1989;**40**:2404-2409

[25] Jones PH, Marag OM, Volpe G. Optical Tweezers: Principles and Applications. United Kingdom, UK: Cambridge University Press; 2015

[26] Wang F, Chen Y, Lina G, Liu L, Cai Y. Complex gaussian representations of partially coherent beams with nonconventional degrees of coherence. Journal of the Optical Society of America A. 2017;**34**:1824-1829

[27] Janousek J, Morizur J-F, Treps N, Lam PK, Harb C, Bachor H-A. Optical entanglement of co-propagating modes. Nature Photonics. 2009;**3**:399-402

Chapter 7

Hidden Markov Models for Pattern Recognition

Majed M. Alwateer, Mahmoud Elmezain, Mohammed Farsi and Elsayed Atlam

Abstract

Hidden Markov Models (HMMs) are the most popular recognition algorithm for pattern recognition. Hidden Markov Models are mathematical representations of the stochastic process, which produces a series of observations based on previously stored data. The statistical approach in HMMs has many benefits, including a robust mathematical foundation, potent learning and decoding techniques, effective sequence handling abilities, and flexible topology for syntax and statistical phonology. The drawbacks stem from the poor model discrimination and irrational assumptions required to build the HMMs theory, specifically the independence of the subsequent feature frames (i.e., input vectors) and the first-order Markov process. The developed algorithms in the HMM-based statistical framework are robust and effective in real-time scenarios. Furthermore, Hidden Markov Models are frequently used in real-world applications to implement gesture recognition and comprehension systems. Every state of the model can only observe one symbol in the Markov chain. In contrast, every state in the topology of a Hidden Markov Model can see one symbol emerging from a particular gesture. The matrix representing the observation probability distribution contains the likelihood of observing a symbol in each state. As an illustration, the probability that a symbol will emit is determined by its observation probability in the first state. In the recognition task, the emission distribution is another name for the observation probability distribution. For the following reasons, HMM states are also referred to as hidden. First, choosing to emit a symbol denotes the second process. Second, an HMM's emitter only releases the observed symbol. Finally, since the current states are derived from the previous states, the emitting states are unknown. HMMs are well-known and more flexible in the field of gesture recognition because of their stochastic nature.

Keywords: recognition algorithm, gesture recognition, hidden Markov models, statistical framework, probability distribution matrix

1. Introduction

One of the greatest methods for pattern recognition is HMM since it can deal with the issues caused by spatiotemporal variability [1–4]. Additionally, HMMs have been effectively used in the classification of gestures, speech, modeling of

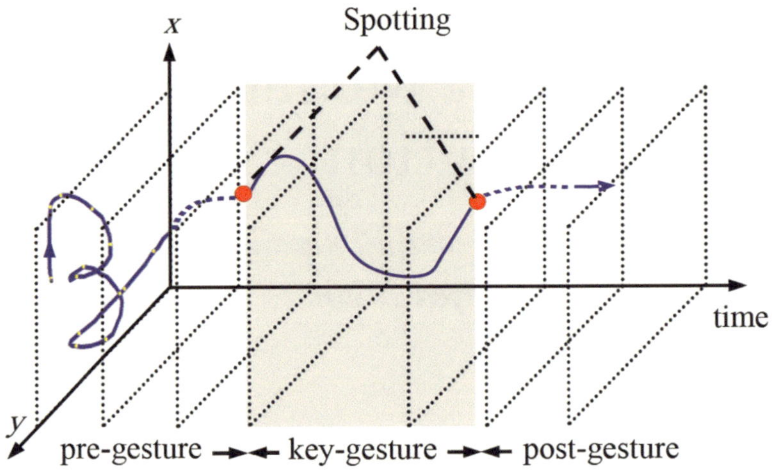

Figure 1.
Three main phases for the trajectory for spotting meaningful gestures.

proteins, etc. [2, 5–8]. Because segmentation and recognition are tuned concurrently during recognition using HMMs, the use of HMMs increases the effectiveness of recognition-based segmentation. The two categories of gesture are communicative gestures (also known as key gestures or meaningful gestures) and noncommunicative gestures (i.e., garbage gesture or transition gesture) [9–12]. In other words, as indicated in **Figure 1**, a natural gesture comprises three phases: pre-, key-, and post-gesture. The key gesture may be described as a hand trajectory component that has clear human significance. While pre- and post-gestures signify inadvertent motion utilized to integrate important gestures (**Figure 1**).

A method to detect American Sign Language was developed by Vogler and Metaxas [13], using HMMs and three video cameras (ASL). To extract 3D characteristics of the user's hand and arm, they employed an electromagnetic tracking device. Two trials using 99 texts and 22 signals to test their method were conducted. Both single signals and full phrases were used in the trials. For isolated signs and continuous phrases, their system has obtained recognition rates of 94.5 and 84.5%, respectively. A data collection of 40 signs was used by Starner *et al.* [5] to present two HMM-based algorithms. While the second technique embeds the camera on the user's headgear, the first system depends on the camera's presence on a desk. According to hand shapes as extracted characteristics, the trials have been run on continuous data collection. The systems accuracy rates for the first and second systems were 92 and 98%, respectively.

A German Sign Language (GSL) recognition system based on HMMs was presented by Bauer and Kraiss [14] and uses colored gloves. Both continuous and isolated indicators have been used to test the system. Subunit HMMs were employed in their system to identify isolated signs and carry out the spotting signs. The system detected spotted hand signals with an accuracy of 92.5% for 100 isolated signs, according to the findings of the experiment.

For French Sign Language (FSL), Braffort [15] presented a recognition method that classified signs into communicative, noncommunicative, and variable signals. In order to extract information like hand position and appearance, a colored glove was utilized. These features were then used to generate HMM codewords. The studies followed classifiers; one was used to identify communicative signs, and the other to identify both noncommunicative and variable indicators. For a repertoire of seven signs, their method achieves a 96% recognition rate.

The approach used by Lee and Kim [6] is regarded as one of the pioneering examples of distinct modeling that deals with transition gestures. Regardless of the hand morphologies, this approach has been utilized to handle 2D hand trajectory (also known as gesture trajectory). The disadvantage of this technique is that while merging two states, the quantity of samples is not taken into account.

The Hierarchical Activity Segmentation (HAS) method was proposed by Kahol *et al.* [16, 17]. The HAS algorithm dynamically represented human anatomy using layers of hierarchy. In addition to making several tries to segment a complicated human motion sequence, this algorithm has also taken low-level motion factors into account to distinguish up-bottom motions (e.g., dancing). There were two primary phases that the mechanism of this procedure went through. Three cues were used in the first step to identify the boundaries of potential gestures, and they were then applied in the second step to a naive Bayesian classifier for boundary detection of the right gesture. Individual gesture patterns were analyzed in 3D using coupled HMMs (cHMMs) in order to identify dance sequences. Unlike other studies, which are only interested in key gesture spotting, this method takes into account all transition gestures in person motions. The transition gestures are explicitly modeled in order to identify key gestures precisely. In summary, HMMs effectively model the spatiotemporal aspects of gestures and can manage non-gesture patterns.

The remainder of the chapter is organized as next. Section 2 explores an overview of the Hidden Markov Model. Section 3 conducts hand gesture recognition application and Section 4 provides the concluding remarks.

2. Hidden Markov model

The most popular method for gesture recognition algorithms is Hidden Markov Models (HMMs) [1, 18–20]. HMMs are mathematical models of the stochastic process and observations, which should be estimated, using a well-defined approach based on previously stored information. The statistical approach in HMMs has numerous benefits as a strong mathematical framework, effective learning and decoding techniques, the ability to handle sequences well, and flexible topology for the statistical syntax. HMMs are a viable option for modeling dynamic gestures since they are time-varying processes that exhibit statistical variation and are frequently employed in practice to put gesture recognition and comprehension systems into place.

Every state in the Markov chain can only see one symbol, whereas the architecture of HMMs allows for more observation, every state can see one sign emerging from a different gesture. The likelihood of finding a symbol for each state is contained in the observation probability distribution matrix. For instance, the probability of observing a symbol in the state s_1 is equivalent to the probability of the symbol emitting. Furthermore, HMMs states are called hidden because of (1) Choosing to observe a symbol denotes the execution of the second process; (2) an HMM emitter only emits the symbol that has been chosen to be seen; and (3) the emitting states are unknown since the current states are derived from the prior states. Due to their stochastic character, HMMs are more adaptable and well-known in the field of gesture recognition.

2.1 Elements of HMMs

HMMs can be expressed with $\lambda = (A, B, \pi)$ and are defined by the following parameters [1, 21–23];

- The group of states $S = \{s_1, s_2, \ldots, s_N\}$. N express the number of states in the technique.

- For each state π, an initial probability distribution that;

$$\pi_i = P(s_i), \quad 1 \leq j \leq N \tag{1}$$

- An N-by-N transition matrix $A = \{a_{ij}\}$, which defined by;

$$a_{ij} = P(s_j | s_i), \quad 1 \leq i, j \leq N \tag{2}$$

where a_{ij} is the likelihood that state s_i at time t will change to state s_j at time $t + 1$. since the probabilities of changing from one state to another are represented by the entries in each row of matrix A their sum must be equal to 1.

$$\sum_j a_{ij} = 1 \tag{3}$$

- The potential observations list $O = \{o_1, o_2, \ldots, o_T\}$ in which T is the gesture path length.

- The group of symbols $V = \{v_1, v_2, \ldots, v_M\}$,

where M denotes the different observation symbols for each state (Size of code word).

An N-by-M observation matrix $B = \{b_j(m)\}$, where

$$b_j(m) = P(v_m | s_j), \quad 1 \leq j \leq N, \quad 1 \leq m \leq M \tag{4}$$

$$\sum_m b_j(m) = 1 \tag{5}$$

where $b_j(m)$ represents the probability that symbol v_m would be emitted at state s_j. For the same obvious reason, the total of each row's elements in matrix B must equal 1.

Two model parameters (N and M) make up a comprehensive specification of the HMMs. It also contains the three probabilistic parameters A, B, and π as well as the observation symbols. Thus, the following is a compact notation for HMM:

$$\lambda = P(\pi, A, B) \tag{6}$$

Here, λ refers to the model's set of parameters.

2.2 HMMs problems

Mathematically, three variables govern the application of HMMs. These elements are topologies, the chosen emission features, and the probabilities of their observations. The observation job informs the feature choices. Moreover, there are three primary problems that must be resolved to effectively use transition probabilities in practical applications for real-world applications. They are:

- **Evaluation problem**: How to determine the probability of the observed sequence given the observation sequence O's model parameter and the model parameter λ (i.e., $P(O|\lambda)$)?

- **Decoding problem**: How to choose the most efficient route λ that generates $O = \{o_1, o_2, \dots, o_T\}$ with maximum probability for observation sequence O and model parameter λ,

- **Estimation problem**: How to change or update the model $\lambda = P(\pi, A, B)$ to generate $O = \{o_1, o_2, \dots, o_T\}$ with largest probability? for observation sequence O.

2.2.1 Evaluation problem

The simple way is to calculate $P(O|\lambda)$ by counting all conceivable state sequences s of length T and calculating the corresponding probability $P(O|s_1, s_2, \dots, s_t)$ for given observation sequence O and model parameter λ. Assume that the definition of the forward chain is equal to the probability of the partial observation sequence $O = o_1, o_2, \dots, o_t$ at state s_i equals the definition of the forward variable $\alpha t(i)$ (**Figure 2**) [1]. Therefore,

$$\alpha_t(i) = P(o_1, o_2, \dots, o_t, s_i|\lambda) \tag{7}$$

Using the following recursive relation, the computation of all forward variables is simple α's

$$\alpha_{t+1}(j) = \sum_{i=1}^{N} \alpha_t(i) \cdot a_{ij} \cdot b_j(o_{t+1}), \quad 1 \le j \le N, \quad 1 \le t \le T-1 \tag{8}$$

where the α values are calculated as follows:

$$\alpha_1(j) = \pi_j \cdot b_j(o_1), \quad 1 \le j \le N \tag{9}$$

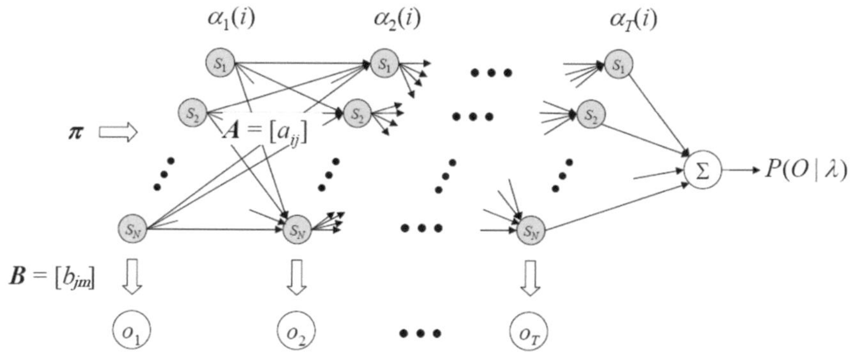

Figure 2.
Forward algorithm for trellis diagram.

Therefore, the required probability $P(O|\lambda)$ with the terminated procedure at T is provided by:

$$P(O|\lambda) = \sum_{i=1}^{N} \alpha_T(i) \tag{10}$$

The probability of the partial observation sequence is similar $o_{t+1}, o_{t+2}, \dots, o_T$ at the state s_i for backward variable $\beta_t(i)$ is defined as:

$$\beta_t(i) = P(o_{t+1}, o_{t+2}, \dots, o_T, s_i|\lambda) \tag{11}$$

Moreover, compute the recursive relationship $\beta_t(i)$ for calculations of α's as follows:

$$\beta_t(i) = \sum_{j=1}^{N} \beta_{t+1}(j) \cdot a_{ij} \cdot b_j(o_{t+1}), \quad 1 \le i \le N, \quad 1 \le t \le T-1 \tag{12}$$

Then,

$$\beta_T(i) = 1, \quad 1 \le i \le N \tag{13}$$

Hence, for state s_i at time t, the multiplication of β and α is calculated as follows:

$$\alpha_t(i) \cdot \beta_t(i) = P(O, s_i|\lambda), \quad 1 \le i \le N, \quad 1 \le t \le T \tag{14}$$

This estimate offers a different calculation $P(O|\lambda)$ using both forward α's and backward β's parameters as in (Eq. (15)).

$$P(O|\lambda) = \sum_{i=1}^{N} \alpha_t(i) \cdot \beta_t(i) \tag{15}$$

Finally, the prior equation is crucial and more precise in predicting the formulas required for gradient-based training.

2.2.2 Decoding issue

To explain the precise state sequence that produces the observations $O = o_1, o_2, \dots, o_t$ with model parameter $\lambda = (A, B, \pi)$ and maximum probability is the incentive for solving the decoding problem. Therefore, the Viterbi algorithm which is applied to find the sequence of most likely hidden states to emit the sequence of observed events is employed as in **Figure 3** [20].

The Viterbi path is a set of hidden states. The forward variable $\alpha_t(i)$ is implemented similarly to the Viterbi algorithm. During the recursion stage, the maximize over the preceding states makes a difference. An auxiliary variable to make computation easier is defined as follows:

$$\delta_t(j) = \max\{P(o_1, o_2, \dots, o_t, s_1, s_2, \dots, s_t|\lambda)\} \tag{16}$$

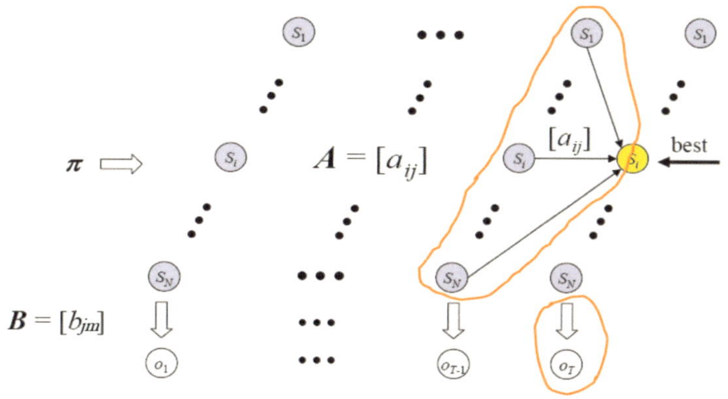

Figure 3.
Forward algorithm for trellis diagram with $\delta_t(j)$ of largest probability of arriving in statej at time t.

To demonstrate how the Viterbi algorithm works, the following steps are used:

- Step1 (Initialization): *for* $1 \le i \le N$,
  (a) $\delta_1(i) = \pi_i \cdot b_i(o_1)$
  (b) $\phi_1(i) = 0$
- Step2 (Recursion): *for* $2 \le t \le T, \ 1 \le j \le N$,
  (a) $\delta_t(j) = \max_i \left[\delta_{t-1}(i) \cdot a_{ij}\right] \cdot b_j(o_t)$
  (b) $\phi_t(j) = \arg\max_i \left[\delta_{t-1}(i) \cdot a_{ij}\right]$
- Step3 (Termination):
  (a) $p^* = \max_i [\delta_T(i)]$
  (b) $q_T^* = arg \max_i [\delta_T(i)]$
- Step4 (Reconstruction): *for* $t = T - 1, \quad T - 2, \dots, 1$

$$q_t^* = \phi_{t+1}(q_{t+1}^*)$$

The resulted optimal states sequence and tate optimized likelihood function is $q_1^*, q_2^*, \dots, q_T^*$. $\phi_t(j)$ indicates the state index j at time t, and p^*.

2.2.3 Estimation problem

The system's effectiveness is greatly impacted by the training procedure. A HMM is trained by modifying its model parameters to get the best description for the observation sequence O_{train}. There is currently no analytical method for optimizing HMMs parameters to optimize the likelihood of observed sequences from the training data set [1]. Instead, the Baum-Welch (BW) method is employed to complete the training procedure, so that $\lambda = (A, B, \pi)$ is optimized with maximum likelihood $P(O|\lambda)$ [24]. BW is a generalized expectation maximization procedure that is calculated using variables that are both forward and backward [25]. Moreover, BW algorithm uses a number of iterations for the observation sequence to optimize HMMs parameters. BW determines the posterior mode estimate and the maximum likelihood estimation for the HMMs parameters (A, B, π) given a series of observation sequences $o_{train} \in O$.

To demonstrate how the BW algorithm works, two auxiliary variables are defined. The probability of traveling along an arc i at time t to state j at time $t+1$ (**Figure 4**) is as follows:

$$\xi_t(i,j) = P(s_i \ at \ t, s_j \ at \ t+1|O,\lambda) \tag{17}$$

where the transition probability is representing by ξ and Eq. (17) becomes as:

$$\xi_t(i,j) = \frac{P(s_i \ at \ t, s_j \ at \ t+1|O,\lambda)}{P(O|\lambda)} \tag{18}$$

However, by using forward and backward variables, Eq. (18) becomes:

$$\xi_t(i,j) = \frac{\alpha_t(i) \cdot a_{ij} \cdot \beta_{t+1}(j) \cdot b_j(o_{t+1})}{\sum_{i=1}^{N}\sum_{j=1}^{N}\alpha_t(i) \cdot a_{ij} \cdot \beta_{t+1}(j) \cdot b_j(o_{t+1})} \tag{19}$$

The second variable (state probability is defined by;

$$\gamma_t(i) = P(s_i \ at \ t|O,\lambda) \tag{20}$$

where $\gamma_t(i)$ is the probability of state i at t for given the model parameters and the observation sequence (**Figure 4**). The following formula Eq. (20) is used to calculate forward and backward variables:

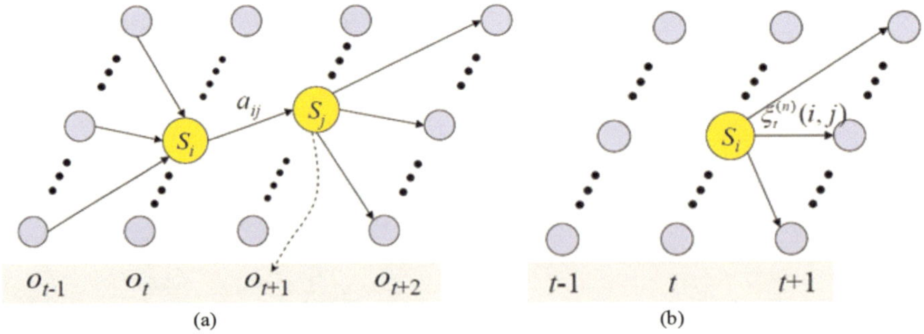

Figure 4.
Trellis diagram learning method according to Baum-Welch. (a) the likelihood of making the journey from state i to state j. (b) the likelihood that state i at time t.

$$\gamma_t(i) = \frac{\alpha_t \cdot \beta_{t+1}}{\sum_{i=1}^{N} \alpha_t \cdot \beta_{t+1}} \tag{21}$$

Therefore, the relationship between $\gamma_t(i)$ and $\xi_t(i,j)$ becomes;

$$\gamma_t(i) = \sum_{i=1}^{N} \xi_t(i,j), \quad 1 \leq i \leq N, \quad 1 \leq t \leq M \tag{22}$$

As a result, BW algorithm modifies the new HMM parameters to have the highest likelihood under the condition $P(O|\lambda)$. The α' and β' can be calculated using the recursive equations of (8) and (12) given the initial parameters $\lambda = (A, B, \pi)$. Eqs. (19) and (22) are used to calculate the auxiliary variables ξ' and γ', respectively. Moreover, the HMMs parameters are updated as follows:

$$\pi' = \gamma_1(i), \quad 1 \leq i \leq N, \tag{23}$$

$$a'_{ij} = \frac{\sum_{t=1}^{T} \xi_t(i,j)}{\sum_{t=1}^{T-1} \gamma_t(i)}, \quad 1 \leq i \leq N, \quad 1 \leq j \leq N, \tag{24}$$

$$b'_j(k) = \frac{\sum_{t=1}^{T} \gamma_t(j) \cdot \zeta_{k,o_t}}{\sum_{t=1}^{T} \gamma_t(j)}, \quad 1 \leq i \leq N, \quad 1 \leq k \leq M, \tag{25}$$

where ζ_{k,o_t} is defined as follows:

$$\zeta_{k,o_t} = \begin{cases} 1 & k = o_t \\ 0 & \text{otherwise} \end{cases} \tag{26}$$

2.3 Topologies of HMMs

Because it depends on the training data that is available and the intended model to represent, HMMs topology has a considerable impact on the success of the recognition process. There are three types of HMMs: (1) Fully connected models have positive a_{ij} Ergodic topology coefficients because every state of the model may be reached from any other state. **Figure 5** displays topology for $N = 4$ the state model.

Therefore, the transition among states has become:

$$A = \begin{pmatrix} a_{11} & a_{12} & a_{13} & a_{14} \\ a_{21} & a_{22} & a_{23} & a_{24} \\ a_{31} & a_{32} & a_{33} & a_{34} \\ a_{41} & a_{42} & a_{43} & a_{44} \end{pmatrix} \tag{27}$$

(2) [18] for the Left–Right model or the Bakis model. According to **Figure 6**, each state of this model can either proceed to itself or one of the following states. The state transition coefficients of this model have the feature that, for a $N = 4$ as illustrated in **Figure 6** (which means state transitions whose indexes are lower than the current state are not allowed);

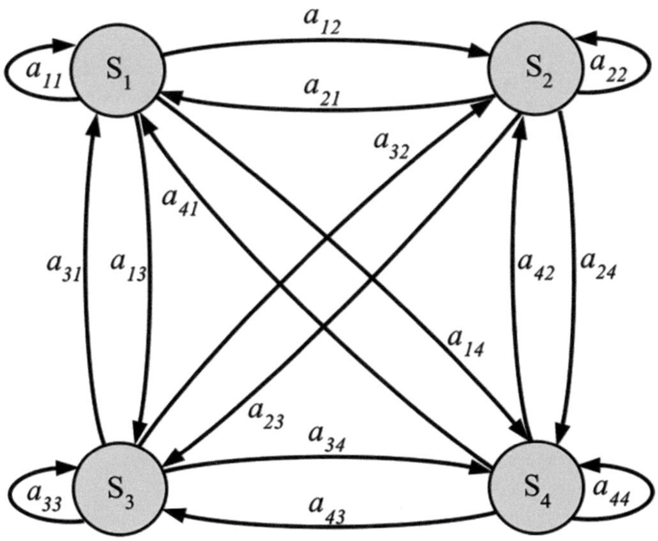

Figure 5.
Ergodic model with four states.

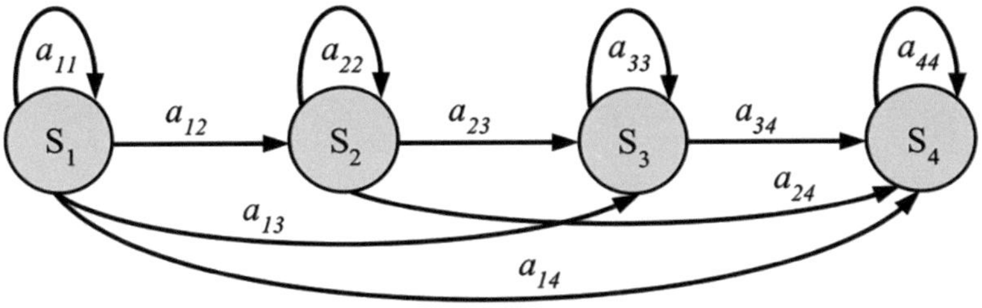

Figure 6.
Left–right model with four states.

$$a_{ij} = 0, \quad j < i \tag{28}$$

Moreover, the model's starting state probabilities possess the characteristic;

$$\pi_i = \begin{cases} 0 & i \neq j \\ 1 & i = j \end{cases} \tag{29}$$

where the state sequence must begin from s_1 and transition among states of **Figure 6** becomes;

$$A = \begin{pmatrix} a_{11} & a_{12} & a_{13} & 0 \\ 0 & a_{22} & a_{23} & a_{24} \\ 0 & 0 & a_{33} & a_{34} \\ 0 & 0 & 0 & a_{44} \end{pmatrix} \tag{30}$$

The transition coefficients, specifically for the final state in a left–right model, are described as follows using the laws of probability:

$$a_{NN} = 1, \quad a_{Ni} = 0, \quad i < N \tag{31}$$

(3) The Left–Right Banded model (LRB) is shown in (7) [21]. The LRB model arranges the states from left to right, and each one has a positive probability of returning to either itself or the one after it. The model's state transition coefficients exhibit the characteristic;

$$A = \begin{pmatrix} a_{11} & a_{12} & 0 & 0 \\ 0 & a_{22} & a_{23} & 0 \\ 0 & 0 & a_{33} & a_{34} \\ 0 & 0 & 0 & a_{44} \end{pmatrix} \tag{32}$$

Therefore, in an LRB model, the last state's and other states' state transition coefficients are as follows:

$$a_{NN} = 1, \quad a_{ij} = 0, \quad i < j; \quad j - i \geq 2 \tag{33}$$

and any HMM parameter with a zero initial value stays zero throughout the estimate procedure. The Left–Right Banded model will thus have a detrimental impact on the process (**Figure** 7).

3. Conducting hand gesture recognition application

A gesture is a spatiotemporal pattern that might be static, dynamic, or both.[1] The term "spatio" refers to locating the hand motion in each frame. To be temporal means

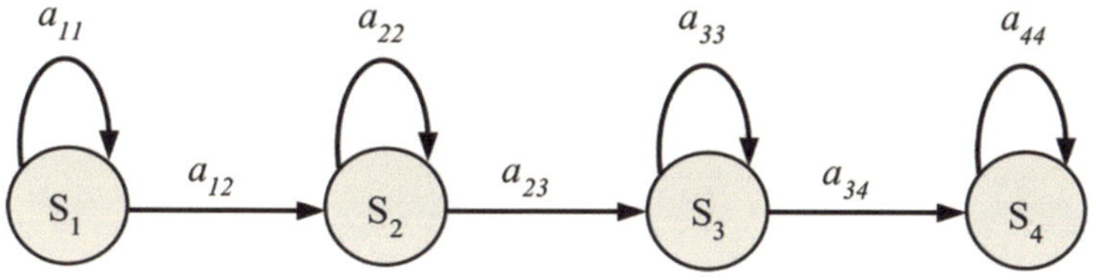

Figure 7.
Left–right banded model with four states.

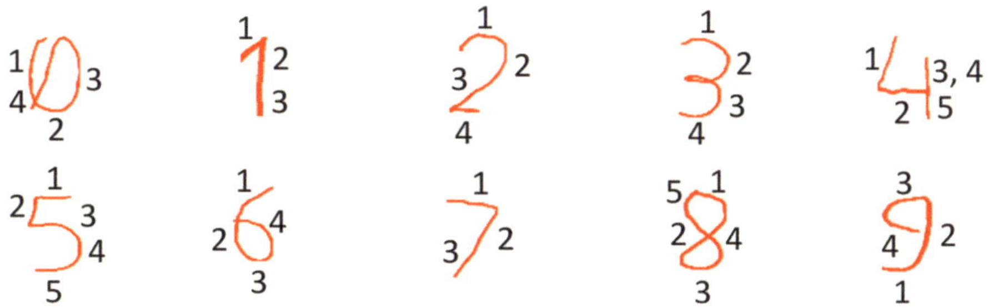

Figure 8.
Hand gesture paths of numbers (0–9) with respect to their segmented parts.

[1] Gestures are hand motions, while postures are static morphs of the hands.

to know when a gesture begins and ends. By comparing the identified hand correspondences between subsequent frames, the hand gesture path is determined (**Figure 8**). It goes without saying that choosing effective features to identify the hand gesture path has a big impact on how well a system performs. The combined features of location, orientation, and velocity with respect to Cartesian systems are used to enhance recognition rates. The HMM codewords are then clustered using K-means. Hand segmentation and feature extraction in computer vision are important steps in developing vision-based hand gesture recognition systems and classification methods for practical applications. Each isolated gesture in our experimental results was based on 60 video sequences, of which 42 were used to train the Baum-Welch algorithm and 18 to test it (i.e., in total, there are 1512 video samples in our database, of which 648 are used for testing). To place a gesture in the appropriate category, the gesture recognition module compares the hand gesture path to a database of reference gestures. The Viterbi algorithm has calculated the greater priority to recognize numbers frame by frame (**Figure 9**). Our application demonstrated good results to identify isolated integers in real-time implementation. The training and testing procedures are covered in the following subsections.

3.1 Model training

The Baum-Welch technique is essential to the proposed system for classifying gestures since it is utilized to train the initialized HMMs parameters $\lambda = (A, B, \pi)$. To place the tested gesture in the appropriate class, the gesture recognition module compares it to a database of reference gestures. As a result, the Viterbi algorithm recognizes the hand gesture path that maximizes the likelihood of all gesture models. According to **Figure 9**, the maximal gesture is the one that has the greatest value out of all the gesture models.

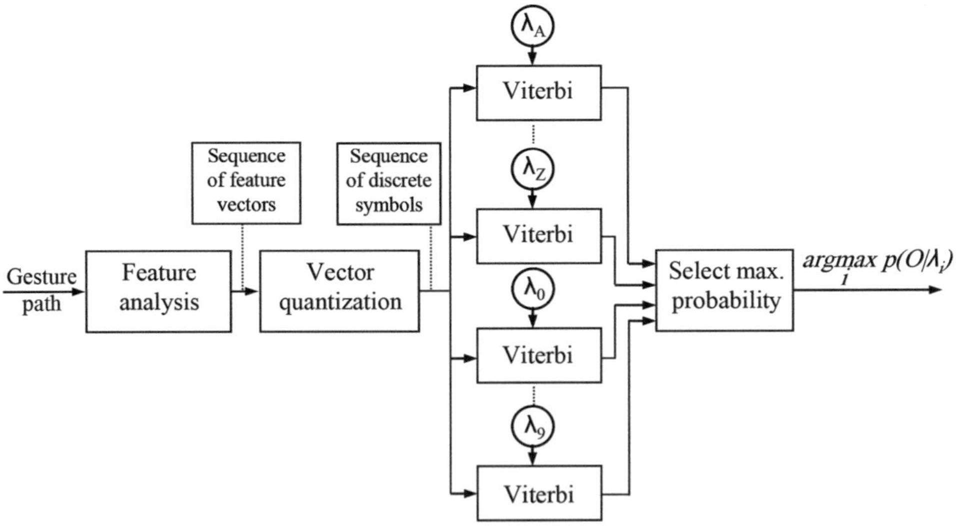

Figure 9.
Block representation of a meaningful gesture recognized by HMMs (Viterbi).

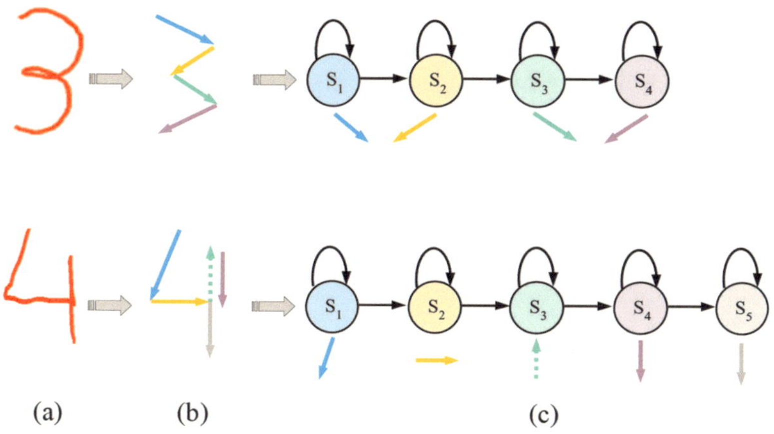

Figure 10.
HMM topologies with a straight-line segment (a) the gesture number derived from the hand motion trajectory (b) the gesture number's line segment (c) LRB model with segmented line codewords.

3.1.1 Model size

The size of the HMMs must be chosen before the HMMs training process begins. We need how many states, exactly? The complexity of the numerous patterns that will be distinguished by HMMs must be taken into account when estimating the number of states. In other words, when we displayed the graphical pattern, the number of segmented portions was taken into account. The over-fitting[2] issue arises when there are insufficient training data samples and excessive state numbers are used.

Additionally, using too few states reduces the discrimination capability of HMMs since several segmented parts of a graphical pattern are modeled using the same state. Our gesture recognition method maps each straight-line segment into a single HMM state, which establishes the number of states (**Figure 10**). We must examine potential patterns to determine how many distinct segments are contained in each pattern in order to represent different graphical patterns. It could be a good idea to employ varying numbers of states in the various HMMs, which were once used to represent various kinds of patterns. For instance, only two states are necessary to represent a graphical pattern "L," whereas six states are needed to represent a graphical pattern "E," and four states are needed to represent a graphical pattern "3."

3.1.2 Initializing a left: Right banded model

The initial values of each parameter in the HMMs must be set before beginning the iterative Baum-Welch process. The initial model must somehow indicate how we want to represent various model states, and this is the only general need.

[2] When HMMs describe random error rather than the underlying relationship, over-fitting takes place. Potential over-fitting issues depend not only on the quantity and quality of data and parameters, but also on how well the model structure fits with the degree of model error and the nature of the data. When more training does not improve generalization, additional strategies such as regularization, early stopping, cross-validation, and others are utilized to prevent the issue of over-fitting. The reader can seek for additional information with [26].

Nevertheless, depending on the HMMs, this condition has a variety of effects. In reality, the LRB model is taken into account because each state in ergodic topology has more transitions than LR and LRB topologies, making it easy to lose structure data. In contrast, because the LRB topology excludes a backward transition, the state index either rise over time or stays constant. LRB topology is also simpler for training data than LR topology and more constrained, allowing for data-to-model matching [21].

It seems intuitive that a proper initialization of the HMMs (A, B, π) parameters leads to better outcomes. The first parameter is found by applying Eq. (34) to Matrix A.

$$A = \begin{pmatrix} a_{11} & 1 - a_{11} & 0 & \cdots & 0 \\ 0 & a_{22} & 1 - a_{22} & \cdots & 0 \\ \vdots & \vdots & \vdots & \ddots & \vdots \\ 0 & 0 & 0 & \cdots & 1 \end{pmatrix} \tag{34}$$

It is possible to select the transition matrix's diagonal elements $a_i i$ to roughly represent the average state durations d in the following ways:

$$a_{ii} = 1 - \frac{1}{d} \tag{35}$$

and

$$d = \frac{T}{N} \tag{36}$$

Such that T denotes the gesture path's length and N is the number of states.

This is adequate for an automatic training method where state 1 is intended to represent the first part of the training data, state 2 the second part, etc. As a result, the same parameters can be used to initialize all output probability distributions for various states. As a result, the training data are used to determine more accurate output probability parameters for each state in the first iteration of the Baum-Welch algorithm. Due to the discrete nature of HMM states, all members of matrix B are initially set to the same value for each state (Eq. (38)). In the matrix B, which consists of N-by-M observed symbols, b_{im} denotes the likelihood that symbol v_m would emit in state i (Eq. (4)).

$$b_{im} = \frac{1}{M} \tag{37}$$

where i, m represent the total number of states and discrete symbols, respectively.

$$B = \begin{pmatrix} b_{11} & b_{12} & \cdots & b_{1M} \\ b_{21} & b_{22} & \cdots & b_{2M} \\ \vdots & \vdots & \ddots & \vdots \\ b_{N1} & b_{N2} & \cdots & b_{NM} \end{pmatrix} = \begin{pmatrix} \frac{1}{M} & \frac{1}{M} & \cdots & \frac{1}{M} \\ \frac{1}{M} & \frac{1}{M} & \cdots & \frac{1}{M} \\ \vdots & \vdots & \ddots & \vdots \\ \frac{1}{M} & \frac{1}{M} & \cdots & \frac{1}{M} \end{pmatrix} \tag{38}$$

The state can return on its own or merely to the next closest state for each new time sample. Consequently, the π initial probability vector should be set up as:

$$\pi = (1 \ 0 \ \cdots \ 0)^T \tag{39}$$

This will guarantee that we begin from the initial state.

3.1.3 Termination of HMMs learning

The Baum-Welch training algorithm is quite effective. After 5–10 iterations, a satisfactory model is frequently found. The trained model must be flexible enough to faithfully represent an entirely new test sequence after training. The training procedure is repeated until the transition and emission matrices change. If the change is smaller than 0.001, that is, tolerance $\varepsilon = 0.001$) as indicated in Eq. (40), or if the maximum number of iterations is reached (i.e., 500), the convergence is satisfied.

$$\sum_{i=1}^{N}\sum_{j=1}^{N} |\hat{a}_{ij} - a_{ij}| + \sum_{j=1}^{N}\sum_{m=1}^{M} |\hat{b}_{jm} - b_{jm}| < \varepsilon \tag{40}$$

The primary purpose of tolerance is to reduce the number of steps the Baum-Welch algorithm needs to take to successfully complete a task. The process is stopped if any one of the next three quantities falls below the tolerance limit. First, a current observation sequence's log-likelihood (O) is generated using the most recent estimates for the values of the transition matrix (A) and observation matrix (B). Second, a modification in the transition matrix A's normalization. Finally, the observation matrix B's normalization was changed.

Keep in mind that the Baum-Welch algorithm required fewer steps to run before terminating as the level of tolerance increased. In actuality, the maximum number of iterations regulates the maximum number of algorithm execution steps. The termination occurs with a warning if the Baum-Welch algorithm runs for 500 iterations before reaching the chosen tolerance value. If this happens, the algorithm should be allowed to run for a greater maximum number of iterations until it meets the specified tolerance before terminating. The provision of enough training data is typically exceedingly challenging. Because of this, some observations may never occur in the tiny collection of training data, even if we may be aware that they may have happened with some small probability. When a discrete HMM is trained on such data, Baum-Welch will assign a *zero* observation probability to some of the matrix's elements. In this situation, a very small nonzero value may be allocated, and the row matrix would need to be renormalized. The matrix representing the transition probabilities might lead to a similar issue. We purposefully assigned several of the transition probability matrix's entries identically *zero* values for a left–right banded HMM. After the Baum-Welch training, these elements still have zero values, and they should keep it that way. After completing the training process, it is crucial to adjust the HMMs' parameters.

3.2 Model testing

The process of classification greatly benefits from the selection of the best HMM topology. The goal of the work is to create HMM topologies with a variety of state numbers in order to select the topology that will produce the best outcomes for an isolated gesture system. An extracted discrete vector features from image sequences are subjected to the application of HMMs with Left–Right Banded (LRB), Left–Right (LR), and Ergodic topologies. These topologies are taken into consideration with a variety of states ranging from 3 to 10. Our gesture recognition system uses the complexity of each gesture number to calculate the number of states by mapping each straight-line segment into a single HMMs state (**Figures 8** and **10**). For two reasons, the number of states is a significant parameter. First, the over-fitting issue is caused by the use of excessive state numbers when there are not enough training data samples. Second, having too few states reduces the discrimination capacity of HMMs since multiple segmented parts of a graphical pattern are modeled on a single state.

In practice, the LRB model with five states is really used in the gesture recognition system in order to guarantee that all of the states are used. The structure data is easily lost in Ergodic topology since each state has more transitions than in LR and LRB topologies. Completely contrary, the LRB topology avoids a backward transition in which the state index either increases or stays as time goes on. LRB topology is also easier to train on data than LR topology and more constrained in its ability to match data to the model. Additionally, the gesture paths "4" and "5" have the most segmented sections, so using 5 states is taken into consideration to make sure all of these parts are utilized. The recognition ratio is the number of recognized gestures correctly to the number of tested gestures (Eq. (41)).

$$Recognition\ \ ratio = \frac{\#\ recognized\ \ gestures}{\#\ test\ \ gestures} \times 100\% \tag{41}$$

The BW method is used to train the HMM topologies and the Viterbi algorithm is used to test them. From **Figure 11**, the LRB performs the best performance, with an average ratio of 97.78% for states ranging from 3 to 10. Additionally, topologies with 4 states in LR and LRB received the greatest recognition. Additionally, compared to LR and Ergodic topologies, LRB topology is always superior. In **Figure 11**, there is not a significant difference between LRB and LR in terms of outcomes, and the results of

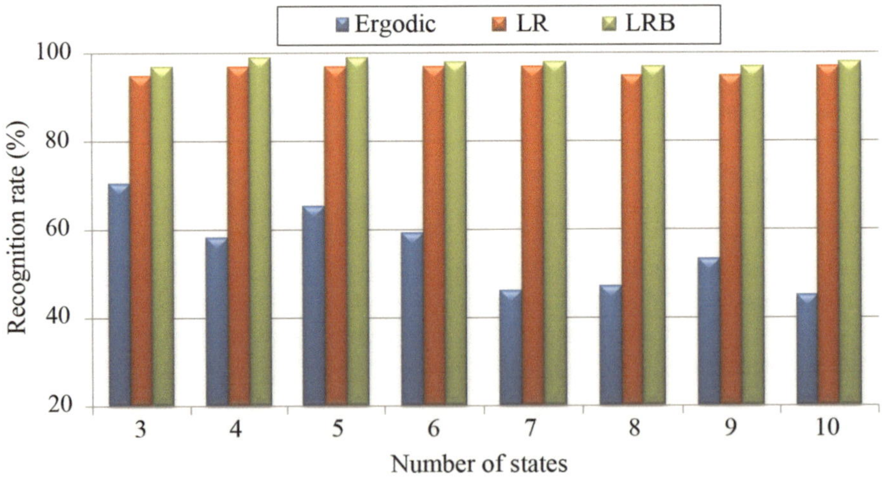

Figure 11.
Results of isolated gesture recognition for three HMMs topologies using states ranging from 3 to 10.

the ergodic topology were subpar in comparison to those of LRB and LR topologies. LRB topology with a number of states equal to 5 is typically the best in terms of its impact on gesture recognition in empirical studies.

4. Conclusions

The foundational Hidden Markov Model method was discussed in this chapter. HMMs are generative models that put a strong emphasis on modeling the observation in order to compute the conditional probability and specify the joint probability distribution in order to resolve a conditional problem. This study assists in determining the best HMM topology and classification method in terms of outcomes. Furthermore, an application was made on recognizing the gesture of the hand for Arabic numbers from 0 to 9 using the classifier of HMMs in terms of their impact on gesture recognition. In addition, studies and research have been conducted on the effects of HMMs with ergodic, left–right, and left–right banded topologies, as well as varying numbers of states ranging from 3 to 10, on hand gesture recognition. It is concluded that there is not a significant difference between LRB and LR in terms of results and that Ergodic topology performed poorly in comparison to LRB and LR topologies. LRB topology with a total of 5 states is typically the best in terms of its effect on gesture recognition.

Abbreviations

HMMs	Hidden Markov Models
ASL	American Sign Language
GSL	German Sign Language
FSL	French Sign Language
HAS	Hierarchical Activity Segmentation
cHMMs	coupled HMMs
BW	Baum-Welch algorithm
LR	Left–Right model
LRB	Left–Right Banded model

Author details

Majed M. Alwateer[1†], Mahmoud Elmezain[1,2*†], Mohammed Farsi[1†] and Elsayed Atlam[1,2†]

1 College of Computer Science and Engineering, Taibah University, Yanbu, Saudi Arabia

2 Computer Science Division, Faculty of Science, Tanta University, Tanta, Egypt

*Address all correspondence to: mahmoud_osman_s@yahoo.com

† These authors contributed equally.

References

[1] Rabiner L. A tutorial on hidden Markov models and selected applications in speech recognition. Proceedings of the IEEE. 1989;**77**(2):257-286

[2] Wang L, Hasegawa-Johnson M. A DNN-HMM-DNN hybrid model for discovering word-like units from spoken captions and image regions. Interspeech. 2020:1456-1460. DOI: 10.21437/interspeech.2020-1148

[3] Wang N, Wang L, Sun Y, Kang H, Zhang D. Three-module modeling for end-to-end spoken language understanding using pre-trained DNN-HMM-based acoustic-phonetic model. Interspeech. 2021:4718-4722. DOI: 10.21437/interspeech.2021-501

[4] Amoolya G, Hans A, Lakkavalli V, Durai S. Automatic speech recognition for Tulu language using Gmm-hmm and DNN-HMM techniques. In: 2022 International Conference on Advanced Computing Technologies and Applications (ICACTA). Coimbatore, India: IEEE; 2022. DOI: 10.1109/ICACTA54488.2022.9753319

[5] Starner T, Weaver J, Pentland A. Real-time American sign language recognition using desk and wearable computer based video. IEEE Transaction on Pattern Analysis and Machine Intelligence. 1998;**20**(12):1371-1375

[6] Lee H, Kim J. An HMM-based threshold model approach for gesture recognition. IEEE Transaction on Pattern Analysis and Machine Intelligence. 1999;**21**(10):961-973

[7] Gao W, Fang G, Zhao D, Chen Y. Transition movement models for large vocabulary continuous sign language recognition. In: IEEE International Conference on Automatic Face and Gesture Recognition. Seoul, Korea (South): IEEE; 2004. pp. 553-558

[8] Yang J, Pan J, Li J. sEMG-based continuous hand gesture recognition using GMM-HMM and threshold model. In: 2017 IEEE International Conference on Robotics and Biomimetics (ROBIO). Macau, Macao: IEEE; 2017. DOI: 10.1109/ROBIO.2017.8324631

[9] Yang H, Park A, Lee S. Gesture spotting and Recognitiomn for human-robot interaction. IEEE Transaction on Robotics. 2007;**23**(2):256-270

[10] Kahol X. Gesture segmentation in complex motion sequences. [Master's thesis] Arizone State University, Tempe, AZ. 2003

[11] Elmezain M, Alwateer M, El-Agamy R, Atlam E, Ibrahim H. Forward hand gesture spotting and prediction using HMM-DNN model. Informatics. 2022;**10**:1. DOI: 10.3390/informatics10010001

[12] Nguyen N, Phan T, Kim S, Yang H, Lee G. 3D skeletal joints-based hand gesture spotting and classification. Applied Sciences. 2021;**11**:4689. DOI: 10.3390/app11104689

[13] Vogler C, Metaxas D. A framework for recognizing the simultaneous aspects of American sign language. Journal of Computer Vision and Image Understanding. 2001;**81**(3):358-384

[14] Bauer B, Kraiss K. Video-based sign recognition using self-organizing subunits. In: International Conference on Pattern Recognition. Quebec City, QC, Canada: IEEE; 2002. pp. 434-437

[15] Braffort A. ARGo: An architecture for sign language recognition and interpretation. In: International Gesture

Workshop Progress in Gestural Interaction. London: IEEE, Springer; 1996. pp. 17-30

[16] Kahol K, Tripath P, Panchanthan S. Automated gesture segmentation from dance sequences. In: IEEE International Conference on Automatic Face and Gesture Recognition. Seoul, Korea (South): IEEE; 2004. pp. 883-888

[17] Kahol K, Tripath P, Panchanthan S. Documenting motion sequences: Development of a personalized annotation system. IEEE Multimedia Magazine. 2006;**13**(1):35-47

[18] Huang X, Ariki Y, Jack M. Hidden Markov Models for Speech Recognition. Taylor & Francis, Ltd.; 1990. DOI: 10.2307/1268779

[19] Elmezain M, Al-Hamadi A, Krell G, El-Etriby S, Michaelis B. Gesture recognition for alphabets from hand motion trajectory using hidden Markov models. In: IEEE International Symposium on Signal Processing and Information Technology. Giza, Egypt: IEEE; 2007. pp. 1192-1197

[20] Elmezain M, Al-Hamadi A, Appenrodt J, Michaelis B. A hidden Markov model-based isolated and meaningful hand gesture recognition. International Journal of Electrical, Computer, and Systems Engineering. Academia education; 2009;**3**(3):156-163 2070-3813

[21] Elmezain M, Al-Hamadi A, Michaelis B. Real-time capable system for hand gesture recognition using hidden Markov models in stereo color image sequences. Journal of WSCG. 2008;**16**(1) ISSN: 1213-6972:65-72

[22] Elmezain M, Al-Hamadi A, Appenrodt J, Michaelis B. A hidden Markov model-based isolated and meaningful hand gesture recognition. In: International Conference on Computer Vision, Image and Signal Processing, PWASET. Vol. 31. Academia.edu; 2008. pp. 394-401

[23] Elmezain M, Al-Hamadi A, Appenrodt J, Michaelis B. A hidden Markov model-based continuous gesture recognition system for hand motion trajectory. In: International Conference on Pattern Recognition. Tampa, FL, USA: IEEE; 2008. pp. 519-522

[24] Baum L, Petrie T, Soules G, Weiss N. A maximization technique occurring in the statistical analysis of probabilistic functions of Markov chains. The Annals of Mathematical Statistics. 1970;**41**(1): 164-171

[25] Soriano M, Huovinen S, Martinkauppi B, Laaksonen M. Skin detection in video under changing illumination conditions. In: Proceeding International Conference on Pattern Recognition. Barcelona, Spain: IEEE; 2000. pp. 839-842

[26] Tetko I, Livingstone D, Luik A. Neural network studies. 1. Comparison of overfitting and overtraining. Journal of Chemical Information and Computer Sciences. 1995;**35**:826-833